DOUGLAS AIRCRAFT COMPANY, INC.

SANTA MONICA PLANT

ENGINEERING DIVISION

presents

PRELIMINARY DESIGN OF AN

EXPERIMENTAL WORLD-CIRCLING SPACESHIP

Report No. SM-11827

Contract W33-038 ac-14105

May 2, 1946

PREPARED BY: F. H. Clauser DOUGLAS AIRCRAFT COMPANY, INC. PAGE: I

DATE: May 2, 1946   SANTA MONICA   PLANT   MODEL: #1033

TITLE: PRELIMINARY DESIGN OF SATELLITE VEHICLE   REPORT NO. SM-11827

SUMMARY

This report presents an engineering analysis of the possibilities of designing a man-made satellite. The questions of power plants, structural weights, multiple stages, optimum design values, trajectories, stability, and landing are considered in detail. The results are used to furnish designs for two proposed vehicles. The first is a four stage rocket using alcohol and liquid oxygen as propellants. The second is a two stage rocket using liquid hydrogen and liquid oxygen as propellants. The latter rocket offers better specific consumption rates, but this is found to be partially offset by the greater structural weight necessitated by the use of hydrogen. It is concluded that modern technology has advanced to a point where it now appears feasible to undertake the design of a satellite vehicle.

FORM 25-S-1
(REV. 8-45)

PREPARED BY: F. H. Clauser   DOUGLAS AIRCRAFT COMPANY, INC.   PAGE: II

DATE: 5-27-46   Santa Monica PLANT   MODEL: #1033

TITLE: PRELIMINARY DESIGN OF SATELLITE VEHICLE   REPORT NO. SM-11827

ABSTRACT

In this report, we have undertaken a conservative and realistic engineering appraisal of the possibilities of building a spaceship which will circle the earth as a satellite. The work has been based on our present state of technological advancement and has not included such possible future developments as atomic energy.

If a vehicle can be accelerated to a speed of about 17,000 m.p.h. and aimed properly, it will revolve on a great circle path above the earth's atmosphere as a new satellite. The centrifugal force will just balance the pull of gravity. Such a vehicle will make a complete circuit of the earth in approximately 1-1/2 hours. Of all the possible orbits, most of them will not pass over the same ground stations on successive circuits because the earth will turn about 1/16 of a turn under the orbit during each circuit. The equator is the only such repeating path and consequently is recommended for early attempts at establishing satellites so that a single set of telemetering stations may be used.

Such a vehicle will undoubtedly prove to be of great military value. However, the present study was centered around a vehicle to be used in obtaining much desired scientific information on cosmic rays, gravitation, geophysics, terrestrial magnetism, astronomy, meteorology, and properties of the upper atmosphere. For this purpose, a payload of 500 lbs. and 20 cu ft. was selected as a reasonable estimate of the requirements for scientific apparatus capable of obtaining results sufficiently far-reaching to make the undertaking worthwhile. It was found necessary to establish the orbit at an altitude of about 300 miles to insure sufficiently low drag so that the vehicle could travel for 10 days or more, without power, before losing satellite speed.

The only type of power plant capable of accelerating a vehicle to a speed of 17,000 m.p.h. on the outer limits of the atmosphere is the rocket. The two most important performance characteristics of a rocket vehicle are the exhaust velocity of the rocket and the ratio of the weight of propellants to the gross weight. Very careful studies were made to establish engineering estimates of the values that can be obtained for these two characteristics.

The study of rocket performance indicated that while liquid hydrogen ranks highest among fuels having large exhaust velocities, its low density, low temperature and wide explosive range cause great trouble in engineering design. On the other hand, alcohol though having a lower exhaust velocity, has the benefit of extensive development in the German V-2. Consequently it was decided to conduct parallel preliminary design studies of vehicles using liquid hydrogen-liquid oxygen and alcohol-liquid oxygen as propellants.

It has been frequently assumed in the past that structural weight ratios become increasingly favorable as rockets increase in size, and fixed weight items such as radio equipment become insignificant weight items. However, the study of weight ratios indicated that for large sizes the weight of tanks and similar items actually become less favorable. Consequently, there is an optimum middle range of sizes. Improvements in weight ratios over that of the German V-2 are possible only by the slow process of technological development, not by the brute force methods of increase in size. This study showed that an alcohol-oxygen vehicle

FORM 25-2-1
(REV. 8-45)

PREPARED BY: F. H. Clauser

DATE: 5-27-46

TITLE: PRELIMINARY DESIGN OF SATELLITE VEHICLE

DOUGLAS AIRCRAFT COMPANY, INC.

Santa Monica PLANT

PAGE: IV

MODEL: #1033

REPORT NO. SM-11827

could be built whose entire structural weight (including motors, controls, etc.) was about 16% of the gross weight. On the other hand, the difficulties with liquid hydrogen, such as increased tank size, necessitated an entire structural weight of about 25% of the gross weight. These studies also indicated that a maximum acceleration of about 6.5 times that of gravity gave the best overall performance for the vehicles considered. If the acceleration is greater, the increased structural design loads increase the structural weight. If the acceleration is less, rocket thrust is inefficiently used to support the weight of the vehicle without producing the desired acceleration.

Using the above results, it was found that neither hydrogen-oxygen nor alcohol-oxygen is capable of accelerating a single unassisted vehicle to orbital speeds. By the use of a multi-stage rocket, these velocities can be attained by vehicles feasible within the limits of our present knowledge. To illustrate the concept of a multi-stage rocket, first consider a vehicle composed of two parts. The primary vehicle, complete with its rocket motor, tanks, propellants and controls is carried along as the "payload" of a similar vehicle of much greater size. The rocket of the large vehicle is used to accelerate the combination to as great a speed as possible, after which, the large vehicle is discarded and the small vehicle accelerates under its own power, adding its velocity increase to that of the large vehicle. By this means we have obtained an effective decrease in the amount of structural weight that must be accelerated to high speeds. This same idea can be used in designing vehicles with a greater number of stages. A careful analysis of the advantages of staging showed that for a given set of performance requirements, an optimum number of stages exists. If the stages are too few in number, the required velocities can be attained only by the undesirable process of exchanging payload for fuel. If they are too many, the multiplication of tanks, motors, etc. eliminates any possible gain in the effective weight ratio. For the alcohol-oxygen rocket it was found that four stages were best. For the hydrogen-oxygen rocket, preliminary analysis indicated that the best choice for the number of stages was two, but refinements showed the optimum number of stages was three. Unfortunately, insufficient time was available to change the design, so the work on the hydrogen-oxygen was completed using two stages. The characteristics of the vehicles studies are tabulated below. Sketches of the vehicles are shown on the drawings preceding page 203.

Vehicle Powered by Alcohol-Oxygen Rockets

Stage	1	2	3	4
Gross Wt. (lbs.)	233,669	53,689	11,829	2,868
Weight less fuel (lbs.)	93,669	21,489	4,729	1,148
Payload (lbs.)	53,689	11,829	2,868	500
Max. Diameter (in.)	157	138	105	90

Vehicle Powered by Hydrogen-Oxygen Rockets

Stage	1	2
Gross Wt. (lbs.)	291,564	15,364
Weight less fuel (lbs.)	84,564	4,464
Payload (lbs.)	15,364	500
Max. Diameter (in.)	248	167

(had three stages been used for the hydrogen-oxygen rockets, the overall gross weight of this vehicle could have been reduced to about 84,000 lbs. indicating this combination should be given serious consideration in any future study).

In arriving at the above design figures, a detailed study was made of the effects of exhaust velocity, structural weight, gravity, drag, acceleration, flight path inclination, and relative size of stages on the performance of the vehicles so that an optimum design could be achieved or reasonable compromises made.

It was found that the vehicle could best be guided during its accelerated flight by mounting control surfaces in the rocket jets and rotating the entire vehicle so that lateral components of the jet thrust could be used to produce the desired control forces. It is planned to fire the rocket vertically upward for several miles and then gradually curve the flight path over in the direction in which it is desired that the vehicle shall travel. In order to establish the vehicle on an orbit at an altitude of about 300 miles without using excessive amounts of control it was found desirable to allow the vehicle to coast without thrust on an extended elliptic arc just preceding the firing of the rocket of the last stage. As the vehicle approaches the summit of this arc, which is at the final altitude, the rocket of the last stage is fired and the vehicle is accelerated so that it becomes a freely revolving satellite.

It was shown that excessive amounts of rocket propellants are required to make corrections if the orbit is incorrectly established in direction or in velocity. Therefore, considerable attention was devoted to the stability and control problem during the acceleration to orbital speeds. It was concluded that the orbit could be established with sufficient precision so that the vehicle would not inadvertently re-enter the atmosphere because of an eccentric orbit.

Once the vehicle has been established on its orbit, the questions arise as to what are the possibilities of damage by meteorites, what temperatures will it experience, and can its orientation in space be controlled? Although the probability of being hit by very small meteorites is great, it was found that by using reasonable thickness plating, adequate protection could be obtained against all meteorites up to a size where the frequency of occurrence was very small. The temperatures of the satellite vehicle will range from about 40°F when it is on the side of the earth facing the sun to about -20°F when it is in the earth's shadow. Either small flywheels or small jets of compressed gas appear to offer feasible methods of controlling the vehicle's orientation after the cessation of rocket thrust.

An investigation was made of the possibility of safely landing the vehicle without allowing it to enter the atmosphere at such great speeds that it would be destroyed by the heat of air resistance. It was found that by the use of wings on the small final vehicle, the rate of descent could be controlled so that the heat would be dissipated by radiation at temperatures the structure could safely withstand. These same wings could be used to land the vehicle on the surface of the earth.

An interesting outcome of the study is that the maximum acceleration and temperatures can be kept within limits which can be safely withstood by a human being. Since the vehicle is not likely to be damaged by meteorites and can be safely brought back to earth, there is good reason to hope that future satellite vehicles will be built to carry human beings.

It has been estimated that to design, construct and launch a satellite vehicle will cost about $150,000,000. Such an undertaking could be accomplished in approximately 5 years time. The launching would probably be made from one of the Pacific islands near the equator. A series of telemetering stations would be established around the equator to obtain the data from the scientific apparatus contained in the vehicle. The first vehicles will probably be allowed to burn up on plunging back into the atmosphere. Later vehicles will be designed so that they can be brought back to earth. Such vehicles can be used either as long range missiles or for carrying human beings.

TABLE OF CONTENTS

FORM 25-5-1
(REV. 8-43)

PREPARED BY: F. H. Clauser DOUGLAS AIRCRAFT COMPANY, INC. PAGE: X

DATE: May 2, 1946 SANTA MONICA PLANT MODEL: #1033

TITLE: PRELIMINARY DESIGN OF SATELLITE VEHICLE REPORT NO. SM-11827

TABLE OF CONTENTS (Cont'd.)

PRELIMINARY DESIGN OF AN
EXPERIMENTAL WORLD-CIRCLING SPACESHIP

1. INTRODUCTION

Technology and experience have now reached the point where it is possible to design and construct craft which can penetrate the atmosphere and achieve sufficient velocity to become satellites of the earth. This statement is documented in this report, which is a design study for a satellite vehicle judiciously based on German experience with V-2, and which relies for its success only on sound engineering development which can logically be expected as a consequence of intensive application to this effort. The craft which would result from such an undertaking would almost certainly do the job of becoming a satellite, but it would clearly be bulky, expensive, and inefficient in terms of the spaceship we shall be able to design after twenty years of intensive work in this field. In making the decision as to whether or not to undertake construction of such a craft now, it is not inappropriate to view our present situation as similar to that in airplanes prior to the flight of the Wright brothers. We can see no more clearly all the utility and implications of spaceships than the Wright brothers could see fleets of B-29's bombing Japan and air transports circling the globe.

Though the crystal ball is cloudy, two things seem clear:

1. A satellite vehicle with appropriate instrumentation can be expected to be one of the most potent scientific tools of the Twentieth Century.

2. The achievement of a satellite craft by the United States would inflame the imagination of mankind, and would probably produce repercussions in the world comparable to the explosion of the atomic bomb.

Chapter 2 of this report attempts to indicate briefly some of the concrete results to be derived from a spaceship which circles the world on a stable orbit.

As the first major activity under contract W33-038AC-14105, we have been asked by the Air Forces to explore the possibilities of making a satellite vehicle, and to present a program which would aid in the development of such a vehicle. Our approach to this task is along two related lines:

1. To undertake a design study which will evaluate the possibility of making a satellite vehicle using known methods of engineering and propulsion.

2. To explore the fields of science in an attempt to discover and to stimulate research and development along lines which will ultimately be of benefit in the design of such a satellite vehicle and which will improve its efficiency or decrease its complexity and cost.

This report concerns itself solely with the first line of approach. It is a practical study based on techniques that we now know. The implications of atomic energy are not considered here. This and other possibilities in the fields of science may be the subject of future reports, which will cover the second line of approach.

In the preliminary design study analytical methods have been developed which may be used as a basis for future studies in this new field of astronautical engineering. Among these are the following:

1. Analysis of single and multi-stage rocket performance and methods for selecting the optimum number of stages for any given application.

2. Dimensional analysis of varying size and gross weight of rockets, deriving laws which are useful in design scaling. These laws are also of assistance in appraisal of the effect of shape and proportions on the design of multi-stage rockets.

3. The effect of acceleration and inclination of the trajectory on structural weight and performance of a satellite rocket.

4. Methods of determining the optimum trajectory for satellite rockets.

5. Variation of rocket performance with altitude and its effect on the proportioning of stages.

6. Preliminary study of effect of atmospheric drag on the rocket and how it affects the choice of stages, acceleration, and trajectory.

7. Analysis of dynamic stability and control throughout the entire trajectory.

8. Method of safely landing a satellite vehicle.

It cannot be emphasized too strongly that the primary contributions of this report are in methods, and not in the specific figures in this design study. One point in particular should be high lighted: - the design gross weight, which is of the greatest importance in estimating cost or in comparing any two proposals in this field is the least definitely ascertained single feature in the whole process. This fact is fundamental in the design of a satellite or spaceship, since the slightest variation in some of the minor details of construction or in propulsive efficiency of the fuel may result in a large change in gross weight. The figures in this report represent a reasonable compromise between the extremes which are possible with the data now in hand. The most important thing is that a satellite vehicle can be made at all in the present state of the art. Even our more conservative engineers agree that it is definitely possible to undertake design and construction now of a vehicle which would become a satellite of the earth.

Another important result of this design study is the conclusion on liquid hydrogen and oxygen as fuel versus liquid oxygen and alcohol (the Germans' fuel). The relative merits of these fuels have occasioned spirited controversy ever since liquid fuel rockets have been under development. In the past, the fact which has clinched the arguments has been the difficulty of handling, storing, and using liquid hydrogen. The present design study has approached this subject from another viewpoint. On the assumption that all these nasty problems can be solved, a design analysis has been made for the structure and performance of rockets using both types of fuels. Because of the low density of liquid hydrogen, the greater tankage weight and volume tends to offset the increase in specific impulse. Early in the design study it was necessary to make a choice of the number of stages for both proposed vehicles. Based on the design information available, a decision was made to use four stages for the alcohol-oxygen rocket and two stages for the hydrogen-oxygen rocket. Of these two designs, the alcohol-oxygen rocket proved to be somewhat smaller in weight and size. However, the problem was later re-examined, when more reliable data were available. It was found that, while the choice of four stages for alcohol-oxygen had been wise, the hydrogen-oxygen rocket could have been substantially improved by using three stages. The improvement was sufficient to indicate that the three stage hydrogen-oxygen rocket would have been definitely superior to the four stage alcohol-oxygen rocket. Unfortunately, the work had progressed so far that it was impossible to alter the number of stages for the hydrogen-oxygen rocket.

One of the most important conclusions of this design study is that in order to achieve the required performance it is necessary to have multi-stage rockets for either type of fuel. The general characteristics of both types are shown in the following table:

4 Stage Alcohol-Oxygen Rocket

Payload 500#

	Stage	1	2	3	4
Gross weight (lbs.)		233,669	53,689	11,829	2868
Fuel weight (lbs.)		140,000	32,200	7,100	1720

2 Stage Hydrogen-Oxygen Rocket

Payload 500#

	Stage 1	2.
Gross weight (lbs.)	291,564	15,364
Total Fuel Weight (lbs.)	217,900	10,000

The design represents a series of compromises. The payload is chosen to be as small as is consistent with carrying enough experimental equipment to achieve significant results. This is done for the purpose of keeping the gross weight within reasonable limits, since the gross weight increases roughly in proportion to the payload above a certain minimum value. The design altitude was originally chosen as 100 miles, since previous calculations indicated that the atmospheric drag there was not great enough to disturb the orbit of the satellite for a few revolutions, and since for communications purposes it was desirable to keep the satellite below the ionosphere. The more refined drag studies made in the present design study show that these early estimates were in serious error, and indicate that the satellite will have to be established at altitudes of 300 to 400 miles to insure the completion of multiple revolutions around the earth.

It is interesting that the design analysis shows that the optimum accelerations are well within the limits which the human body can stand. Further, it appears possible to achieve a safe landing with the type of vehicle which is required. Future developments may bring an increase in payload and decrease in gross weight, sufficient to produce a large manned spaceship able to accomplish important things in a scientific and military way.

We turn now from the design study phase to the basic research approach of the scientists. Our consultants have all made suggestions which have been taken into consideration in the preparation of this report. In the future it is our expectation that the services of these scientists will be of the greatest benefit in planning and initiating broad research programs to explore new fundamental approaches to the problem of space travel.

The real white hope for the future of spaceships is, of course, atomic energy. If this intense source of energy can be harnessed for rocket propulsion, then spaceships of moderate size and high performance may become a reality, and conceivably could even serve efficiently as intercontinental transports in the remote future. We are fortunate in having the consulting services of Drs. Alvarez, McMillan, and Ridenour, well known in scientific circles. Alvarez and McMillan were two of the key men at the Los Alamos Laboratory of the Manhattan Project. With the benefit of their advice, we hope to achieve a degree of competence in the fields of application of nuclear energy to propulsion.

Alvarez and Ridenour, who are also radar experts, have made basic analyses of the radio and radar problems associated with a satellite. These are of service in planning the new equipment which seems to be necessary to make the satellite a useful tool.

Kistiakowsky, a specialist in physical chemistry, has made valuable suggestions for the development of new rocket propellants.

Schiff has contributed to our knowledge of the optimum trajectories to be used in launching the vehicle.

More important than the ideas and suggestions received to date is the fact that these consultants, who are among the leaders in U.S. science, have begun to think and work on these problems. It is our earnest hope that under the terms of this new study and research contract with the Army Air Forces we may be able to enlist the active cooperation of an important fraction of the scientific resources of the country to solve problems in the wholly new fields which man's imagination has opened. Of these, space travel is one of the most important and challenging.

Chapter 2

2. THE SIGNIFICANCE OF A SATELLITE VEHICLE

Attempting in early 1946 to estimate the values to be derived from a development program aimed at the establishment of a satellite circling the earth above the atmosphere is as difficult as it would have been, some years before the Wright brothers flew at Kitty Hawk, to visualize the current uses of aviation in war and in peace. Some of the fields in which important results are to be expected are obvious; others, which may include some of the most important, will certainly be overlooked because of the novelty of the undertaking. The following considerations assume the future development of a satellite with large payload. Only a portion of these may be accomplished by the satellite described in the design study of this report.

The Military Importance of a Satellite - The military importance of establishing vehicles in satellite orbits arises largely from the circumstance that defenses against airborne attack are rapidly improving. Modern radar will detect aircraft at distances up to a few hundred miles, and can give continuous, precise data on their position. Anti-aircraft artillery and anti-aircraft guided missiles are able to engage such vehicles at considerable range, and the proximity fuze increases several fold the effectiveness of anti-aircraft fire. Under these circumstances, a considerable premium is put on high missile velocity, to increase the difficulty of interception.

This being so, we can assume that an air offensive of the future will be carried out largely or altogether by high-speed pilotless missiles. The minimum-energy trajectory for such a space-missile without

Chapter 2

aerodynamic lift at long range is very flat, intersecting the earth at a
shallow angle. This means that small errors in the trajectory of such a
missile will produce large range errors in the point of impact. It has
been suggested that the accuracy can be increased by firing such a missile along the same general course as that being followed by a satellite,
and at such a time that the two are close to one another at the center
of the trajectory of the missile. Under these circumstances, precise
observations of the position of the missile can be made from the satellite, and a final control impulse applied to bring the missile down on
its intended target. This scheme, while it involves considerable complexity in instrumentation, seems entirely feasible. Alternatively, the
satellite itself can be considered as the missile. After observations
of its trajectory, a control impulse can be applied in such direction
and amount, and at such a time, that the satellite is brought down on
its target.

There is little difference in design and performance between an
intercontinental rocket missile and a satellite. Thus a rocket missile
with a free space-trajectory of 6,000 miles requires a minimum energy
of launching which corresponds to an initial velocity of 4.4 miles per
second, while a satellite requires 5.1. Consequently the development of
a satellite will be directly applicable to the development of an intercontinental rocket missile.

It should also be remarked that the satellite offers an observation
aircraft which cannot be brought down by an enemy who has not mastered
similar techniques. In fact, a simple computation from the radar

FORM 25-S-1
(REV. 8-43)
PREPARED BY: L. N. RIDENOUR   DOUGLAS AIRCRAFT COMPANY, INC.   PAGE:
DATE: May 2, 1946   SANTA MONICA   PLANT   MODEL: #1033
TITLE: PRELIMINARY DESIGN OF SATELLITE VEHICLE   REPORT NO. SM-11827

Chapter 2

equation shows that such a satellite is virtually undetectable from the ground by means of present-day radar. Perhaps the two most important classes of observation which can be made from such a satellite are the spotting of the points of impact of bombs launched by us, and the observation of weather conditions over enemy territory. As remarked below, short-range weather forecasting anywhere in the vicinity of the orbit of the satellite is extremely simple.

Certainly the full military usefulness of this technique cannot be evaluated today. There are doubtless many important possibilities which will be revealed only as work on the project proceeds.

The Satellite as an Aid to Research - The usefulness of a satellite in scientific research is very great. Typical of the outstanding problems which it can help to attack are the following:

One of the fastest-moving fields of investigation in modern nuclear physics is the study of cosmic rays. Even at the highest altitudes which have been reached with unmanned sounding balloons, a considerable depth of atmosphere has been traversed by the cosmic rays before their observation. On board such a satellite, the primary cosmic rays could be studied without the complications which arise within the atmosphere. From this study may come more important clues to unleashing the energy of the atomic nucleus.

Studies of gravitation with precision hitherto impossible may be made. This is possible because for the first time in history, a satellite would provide an acceleration-free laboratory where the ever present pull of the earth's gravitational field is cancelled by the centrifugal force

Chapter 2

of the rotating satellite. Such studies might lead to an understanding of the cause of gravitation - which is now the greatest riddle of physics.

The variations in the earth's gravitational field over the face of the earth could be measured from a satellite. This would supply one very fundamental set of data needed by the geologists and geophysicists to understand the causes of mountain-building, etc.

Similarly, the variations in the earth's magnetic field could be measured with a completeness and rapidity hitherto impossible.

The satellite laboratory could undertake comprehensive research at the low pressures of space. The value of this in comparison with pressures now attainable in the laboratory might be great.

For the astronomer, a satellite would provide great assistance. Dr. Shapley, director of the Harvard Observatory has expressed the view that measurements of the ultra-violet spectrum of the sun and stars would contribute greatly to an understanding of the source of the sun's surface energy, and perhaps would help explain sunspots. He also looks forward to the satellite observatory to provide an explanation for the "light of the night sky."

Astronomical observations made on the surface of the earth are seriously hampered by difficulties of "seeing," which arise because of variations in the refractive index of the column of air through which any terrestrial telescope must view the heavens. These difficulties are greatest in connection with the observation of any celestial body whose image is an actual disk, within which features of structure can be

Chapter __2__

recognized: the moon, the sun, the planets, and certain nebulae. A telescope even of modest size could, at a point outside the earth's atmosphere, make observations on such bodies which would be superior to those now made with the largest terrestrial telescopes. Because there would be no scattering of light by an atmosphere, continuous observation of the solar corona and the solar prominences should also be possible. Astronomical images could, of course, be sent back to the earth from an unmanned satellite by television means.

From a satellite at an altitude of hundreds of miles, circling the earth in a period of about one and one half hours, observations of the cloud patterns on the earth, and of their changes with time, could be made with great ease and convenience. This information should be of extreme value in connection with short-range weather forecasting, and tabulation of such data over a period of time might prove extremely valuable to long-range weather forecasting. A satellite on a North-South orbit could observe the whole surface of the world once a day, and entirely in the daylight.

The properties of the ionosphere could be studied in a new way from such a satellite. Present ionospheric measurements are all made by studying the reflection of radio waves from the ionized upper atmosphere. A satellite would permit these measurements to be extended by studying the transmission properties of the ionosphere at various frequencies, angles of incidence, and times. Reflection measurements could also be made from the top of the ionosphere. Since we now know that disruption of the ionosphere accompanying auroral displays is caused by the impact

Chapter 2

of a cloud of matter from space, the satellite could determine the nature, and maybe the source of that cloud.

Biologists and medical scientists would want to study life in the acceleration-free environment of the satellite. This is an important pre-requisite to space travel by man, and it may also lead to important new observations in lower forms of life.

The Satellite as a Communications Relay Station - Long-range radio communication, except at extremely low frequencies (of the order of a few kc/sec), is based entirely on the reflection of radio waves from the ionosphere. Since the properties of the earth's ionized layer vary profoundly with the time of day, the season, sunspot activity, and other factors, it is difficult to maintain reliable long-range communication by means of radio. A satellite offers the possibility of establishing a relay station above the earth, through which long-range communications can be maintained independent of any except geometrical factors.

The enormous bandwidths attainable at microwave frequencies enable a very large number of independent channels to be handled with simple equipment, and the only difficulty which the scheme appears to offer is that a low-altitude (300 mile) satellite would remain in the view of a single ground station only for about 2100 miles of its orbit.

For communications purposes it would be desirable to operate the satellites at an altitude greater than 300 miles. If they could be at such an altitude (approximately 25,000 miles) that their rotational period was the same as that of the earth, not only would the "shadow" effect of the earth be greatly reduced, but also a given relay station could be associated with a given communication terminus on the earth, so that the communication system problem might be very greatly simplified.

Chapter 2

An idea of the potential commercial importance of this development

may be gained from the fact that the ionosphere is now used as the equiva-

lent of about $10,000,000,000. in long-lines, and is jammed to the limit

with transmissions.

Chapter 2

The Satellite as a Forerunner of Interplanetary Travel - The most fascinating aspect of successfully launching a satellite would be the pulse quickening stimulation it would give to considerations of inter-planetary travel. Whose imagination is not fired by the possibility of voyaging out beyond the limits of our earth, traveling to the Moon, to Venus and Mars? Such thoughts when put on paper now seem like idle fancy. But, a man-made satellite, circling our globe beyond the limits of the atmosphere is the first step. The other necessary steps would surely follow in rapid succession. Who would be so bold as to say that this might not come within our time?

Chapter 3

3. GENERAL CHARACTERISTICS OF A SATELLITE VEHICLE

Within the limits of our everyday experience, the trajectories of freely moving objects are nearly parabolic. The departures from truly parabolic trajectories are caused largely by air resistance. However, there is an additional factor whose influence is small at low speeds but rapidly becomes larger as the speed increases. This factor is the curvature of the earth. Because of it, even a vehicle traveling parallel to the earth is subjected to a centrifugal force and at high speeds this force can become of equal importance to the force of gravity. Since gravitational force is inward and the centrifugal force is outward, there is a speed at which the two would just balance and the vehicle would revolve about the earth like a new satellite. The speed to accomplish this is easily calculated. If, for the moment, we disregard aerodynamic forces, then a satellite near the surface of the earth would be balanced between a gravitational attraction of mg and a centrifugal force of $\frac{mv^2}{R}$, where m and v are the mass and velocity of the satellite and g and R are the acceleration of gravity and the radius of the earth. Placing $\frac{mv^2}{R} = mg$ and using the equatorial values of R = 3,963 miles and g = 32.086 ft. per sec.^{2}, we readily find that v = 25,810 ft. per sec. or 17,600 miles per hour. If this motion were to take place in the plane of the equator we would have to add or subtract the velocity of rotation of the earth, depending on whether the vehicle were rotating with or against the earth. These new values are 24,285 ft./sec. and 27,335 ft./sec. These values are only approximately correct because the effect of the earth's rotation on the gravitational attraction of stationary objects has been neglected.

Chapter 3

A more detailed calculation is given in Appendix C. Traveling at these speeds, the times to make a complete circuit of the earth would be 1 hour and 30 min. and 1 hour and 20 min. respectively. It is of course impractical to attempt to move at such great speeds within the atmosphere of the earth. However, at a height of 300 miles above the surface of the earth, the air is so thin that such speeds are practical. If we repeat our calculations for this altitude, taking into account that the attraction of gravity falls off as the square of the distance from the center of the earth, we find that the new velocities are 23,655 ft. per sec. and 26,705 ft. per sec. and the new times for complete circuits of the earth are 1 hour and 32 min. and 1 hour and 22 min. Interestingly enough, the energy required to establish an orbit at an altitude of 200 miles is not very much larger than that required at the surface of the earth because, although the potential energy is considerably greater, it is partly compensated by the lower kinetic energy of the higher orbit.

It is interesting to note that in our equation $\frac{mv^2}{R} = mg$ the mass occurs on both sides and cancels out. Consequently, the speeds for orbital motion do not depend on the mass of the object nor on the material from which it is made.

As mentioned above, we normally expect the trajectories of freely moving objects to be parabolic. However, if we take strict account of the curvature of the earth, our mathematics tells us that all such trajectories are arcs of

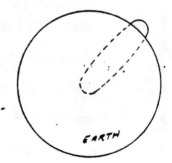

EARTH

ORM 25-S-1
REV. 4-43

PREPARED BY: F. H. Clauser DOUGLAS AIRCRAFT COMPANY, INC. PAGE: 19

DATE: May 2, 1946 (Corr. 5-24-46) SANTA MONICA PLANT MODEL: #1003

TITLE: PRELIMINARY STUDY OF SATELLITE VEHICLE REPORT NO. SM-11827

Chapter 3

Keplerian[*] ellipses. If the velocity is small, the trajectory is only
a small portion of the outer end of the ellipse, as shown in the preceding figure. This tip portion of an elongated ellipse is very nearly but
not quite parabolic. As the speed increases, the portion of the ellipse
lying outside the earth likewise increases and the first trajectory lying
entirely outside the earth is the circular one whose speed was computed
above. As the speed increases still further, the orbits will become

ellipses extending far out into
space as shown in the figure at
the right. Our own moon is, of
course, traveling in an orbit
that is very nearly circular.

So far, only the effect of
velocity on the orbit has been mentioned. However, there is another
factor of importance in determining the characteristics of an orbit,
namely the initial direction with which the body was launched. This, in
turn, will determine whether the orbit is a long flattened ellipse or a
nearly circular one. Both kinds can correspond to the same velocity of
launching, differing only with the direction of launching.

Suppose now that our satellite, mentioned above, is launched directly
upward with the same velocity, instead of on a circular orbit parallel to
the surface of the earth. The simple equation of its motion shows that it
will travel out into space a distance equal to the diameter of the earth
before returning to the earth. If the velocity is increased, the vehicle

[*]After the noted astronomer, Johannes Kepler (d. 1630)

Chapter 3

will, of course, travel even farther out. When the initial velocity has been increased to a value equal to the $\sqrt{2}$ (or 1.41) times the orbital speed of 25,810 ft. per sec. or 36,500 ft. per sec., it will travel out beyond the influence of our planet and never return. This speed is appropriately called the escape velocity.

Returning now to a more detailed examination of the characteristics of a vehicle rotating in a circular orbit a few hundred miles above the surface of the earth, we note that the balance between gravitational and centrifugal forces exists not only for the vehicle itself, but also for all objects within the vehicle. Consequently there will be no "up" or "down". Everything will float weightless inside the vehicle.

When we consider the possible orbits in which the vehicle could travel, as seen from the earth, we realize that they must all be great circle paths, i.e. in planes passing through the center of the earth. Of all such paths, only the one lying in the plane of the equator will repeat itself on each revolution because for all the others when the vehicle has completed a circuit in approximately 1-1/2 hours, the earth has turned under it 1/16 of a revolution and the vehicle is over a new spot on the earth's surface. Consequently, the first attempts at establishing satellites will be around the equator so that they may be repeatedly observed from fixed ground stations.

So far, we have purposely avoided considering the means of supplying the enormous energies necessary to obtain the speeds calculated above. This is such an important problem that it will be given special consideration in the next chapter.

ORM 25-S-1
(REV. 8-43)

PREPARED BY: G. H. Peebles
F. H. Clauser   DOUGLAS AIRCRAFT COMPANY, INC.   PAGE: 21

DATE: May 2, 1946 (Corr. 5-24-6) SANTA MONICA _____PLANT   MODEL: #1003

TITLE: PRELIMINARY DESIGN OF SATELLITE VEHICLE   REPORT NO. SM-11827

Chapter 4

4. POWER PLANT SUITABLE FOR SATELLITE VEHICLES

In order to be able to establish a vehicle in a satellite orbit, a power plant must be capable, not only of lifting its own weight and that of its fuel and the associated structure and payload, but also to accelerate these components sufficiently to attain the enormous velocities calculated in the preceding chapter. Clearly this will require a power plant capable of producing thrusts many times its own weight. At the present time, the only quasi-conventional power plants that meet this requirement are the rocket, the turbo-jet and the ram-jet.

The turbo-jet and the ram-jet both depend upon atmospheric air for their combustion. Their maximum thrusts fall off rapidly with altitude so that their useful range is well below 100,000 ft. When speeds of the order of 24,500 ft. per sec. (Approximately a Mach number of 25) are considered, the compression and friction of the air give calculated temperatures of the order of 49,000°F.* Even at 100,000 ft. the density of the air is sufficiently great to burn up the vehicle in short order. Consequently it would appear that the turbo-jet or the ram-jet could be used only in the very initial stages of launching man-created satellites.

It is conceivable that these power plants may be found to serve a useful purpose as initial launching engines. However, for the present investigation, this scheme has been left out of consideration in order to avoid the complication.

---

*Long before such temperatures are reached, the conventional methods of calculation become invalid. However, the conclusion that the temperatures are prohibitively high is still valid.

FORM 25-S-1 (REV. 8-43)

PREPARED BY: G. H. Peebles, F. H. Clauser

DOUGLAS AIRCRAFT COMPANY, INC.

PAGE: 22

DATE: May 2, 1946

SANTA MONICA PLANT

MODEL: #1003

TITLE: PRELIMINARY DESIGN OF SATELLITE VEHICLE

REPORT NO. SM-11827

Chapter 4

The rocket motor, carrying its own propellants, can traverse the atmosphere at limited speeds and after entering the rarefied ionosphere be free to accelerate to the speeds required for orbital motion. The V-2 has demonstrated the practicality of such a scheme. The greatest question to be answered is whether within the stern accounting of engineering reality, successors to the V-2 can be built capable of accelerating to speeds of the order of 25,000 ft. per sec.

Before attempting to answer this question, it will be of interest to examine rocket power plants in some detail. At present these power plants can be divided into two general classes. The first is the familiar solid propellant type of rocket used extensively in Fourth of July celebrations. When used to obtain high performance, the propellant containers must withstand such great pressures that their weight becomes prohibitive where weight is an important consideration. This has led to the development of the liquid propellent rocket in which the propellants are forced into the combustion chamber under gas pressure (frequently compressed nitrogen) or by means of pumps as in the V-2. For installations where large thrusts are required this latter system has proved to be of lighter weight.

It is helpful to have an understanding of the parameters which are used for evaluating the performance of a rocket motor and which, since they are unique to the field of jet propulsion, may be unfamiliar to the reader. From Newton's familar second and third laws, it may readily be shown that the thrust T is equal to the product of the exhaust velocity, c and the mass rate of propellant consumption, $\frac{dm}{dt}$, thus $T = c \frac{dm}{dt}$. The quantity $\frac{dm}{dt}$ may be made as large as we please since it is only a matter of arranging

ORM 25-5-1
(REV. 8-43)

PREPARED BY: G. H. Peebles
F. H. Clauser

DOUGLAS AIRCRAFT COMPANY, INC.

PAGE: 23

DATE: May 2, 1946

SANTA MONICA   PLANT

MODEL: #1003

TITLE: PRELIMINARY DESIGN OF SATELLITE VEHICLE

REPORT NO. SM-11827

Chapter 4

adequate means for delivering and burning the desired amounts of propellants. However, this is not the case with the exhaust velocity which is more strictly a characteristic of the propellants used. The exhaust velocity is determined to a large extent by the molecular weight, the temperature, and the specific heats of the combustion products. For a given fuel we have little control over these quantities. The pressure in the combustion chamber, the external atmospheric pressure and the overall efficiency of the power plant (which are the factors over which we have greatest control) also affect the exhaust velocity but to such a lesser degree that it is possible to assess the exhaust velocities of an installation largely from a knowledge of the propellants used.

It will be seen later that the exhaust velocity of a rocket installation is of prime importance in determining its suitability for use as a satellite-producing power plant. In addition to the exhaust velocity c, two other parameters are frequently used. The equation for the thrust of a rocket motor shows that a given quantity of propellants, if consumed under comparable conditions, represents an ability to produce a given impulse, either as a large thrust for a short time or a small thrust for a proportionately larger time. Consequently, it is in order to ask for the pounds of thrust obtained per pound of propellant per second. It is seen at once that this parameter, known as the specific impulse I, is given by the formula $I = \dfrac{T}{g\frac{dm}{dt}} = \dfrac{c}{g}$, i.e. it is obtained from the exhaust velocity simply by division by the acceleration of gravity (I is the same in both c.g.s. and ft.-lb.-sec. systems since it contains units of force

ORM 25-5-1
(REV. 8-43)

PREPARED BY: G. H. Peebles
F. H. Clauser

DOUGLAS AIRCRAFT COMPANY, INC.

PAGE: 24

DATE: May 2, 1946   SANTA MONICA   PLANT   MODEL: #1003

TITLE: PRELIMINARY DESIGN OF SATELLITE VEHICLE   REPORT NO. SM-11827

Chapter 4

in both numerator and denominator). Again, we may ask for the pounds of propellant consumed per pound of thrust per second. This is known as the specific fuel consumption and is merely the reciprocal of the impulse: s.f.c. $= \frac{1}{I} = \frac{g}{c}$. Typical values of these parameters are $c = 6,434$ ft. per sec., $I = 200$ sec., and s.f.c. $= .005$ sec.$^{-1}$

In the discussion above, it was mentioned that external atmospheric pressure played a lesser role in determining the exhaust velocity. While this role is small, it is not insignificant and enters into the problem of establishing a man-made satellite in a very helpful fashion. As we go to higher altitudes, the atmosphere exerts less of a back pressure on the exhausting gases, allowing their velocity to increase until at extreme altitudes it has increased by some 20% to 30%. This will be found to be of significant magnitude in our problem of determining if rocket motors are capable of imparting a sufficiently large momentum to the proposed satellite vehicle, a problem to which we now return.

PREPARED BY: P. A. Lagerstrom
F. H. Clauser DOUGLAS AIRCRAFT COMPANY, INC. PAGE: 25

FORM 25-S-1
(REV. 8-43)

DATE: May 2, 1946   SANTA MONICA   PLANT   MODEL: #1003

TITLE: PRELIMINARY DESIGN OF SATELLITE VEHICLE   REPORT NO. #SM-11827

Chapter 5

5. DYNAMICS OF ACHIEVING ORBITAL MOTION

Let us begin by considering the simplest possible case of a rocket motor accelerating our vehicle to high speed. We shall temporarily neglect gravity and air resistance in order to determine what are the fundamental factors occuring in our problem. If m is the mass of the vehicle at any instant, $\frac{dV}{dt}$ the acceleration and T the thrust, then $m\frac{dV}{dt} = T$. In the preceding chapter, we saw that $T = -c\frac{dm}{dt}$.* Placing this in our equation, we have $m\frac{dV}{dt} = -c\frac{dm}{dt}$. This can be integrated to give $\triangle V = c \log \frac{m_0}{m_1}$ where $\triangle V$ is the change in velocity of the vehicle that the rocket produces and m_0 and m_1 are the masses at the beginning and end of the acceleration, their difference being the fuel used in the process. This formula, although it will be successively modified numerous times, brings into focus the two most fundamental parameters of our problem; namely, the exhaust velocity and the mass ratio. In fact, these two parameters are so vital that the next two chapters will be devoted entirely to a critical engineering analysis of what values we can reasonably expect to achieve.

It is clear that the gain in velocity of the vehicle is directly proportional to the exhaust velocity and any improvements in this factor will be immediately reflected in the performance of the vehicle. The mass ratio, entering the logarithm would appear to be a factor of minor importance. However, this appearance is quite deceptive as we shall presently see.

If we put W equal to the initial gross weight of the vehicle, P equal to the payload and S equal to the entire structural, power plant, tank and control weight (i.e. S includes all items except the fuel and the payload)

*The minus sign is necessary here because $\frac{dm}{dt}$ is the rate of change of mass of the vehicle (which is negative) while in Chapter 4 it was the rate of propellant consumption.

Chapter 5

then our formula becomes $\Delta V = -c \ln\left(\frac{S}{W} + \frac{P}{W}\right)$. As we have seen before, c cannot be made arbitrarily large, but is limited by the state of development of our technology. Likewise, $\frac{S}{W}$, the ratio of the entire structural weight to the gross weight, cannot be chosen arbitrarily small but is limited by technological progress. Consequently, the quantity within the parentheses of the logarithm has a smallest value when the payload is zero (this will make the logarithm, with a negative sign in front, have its greatest value). Actually, in engineering application we usually must view this the other way around; that is, the payload is given and the gross weight must be varied. This has been illustrated on the accompanying graph. Here we have plotted the ratio of velocity increase to exhaust velocity against the ratio of gross weight to payload (i.e. the gross weight for a 1 lb. payload) for various values of the structural weight ratio. The extreme importance of this latter parameter is immediately apparent.

This graph also illustrates another characteristic that will confront us time and again; namely, the extreme variability of the gross weight for a fixed payload when we attempt to obtain high performances. For example, suppose we could obtain an exhaust velocity of 11,000 ft. per sec. and build the complete structure for only 5% of the gross weight. Then to accelerate a payload of 1 lb. to 24,750 ft. per sec. ($\frac{\Delta V}{c} = 2.25$) would require a vehicle having a gross weight of 200 lbs. However, if the desired velocity had been only 2% greater, the smallest vehicle with which

---

†Here and throughout the rest of the report we shall refer to both the fuel and the oxidizer simply as the fuel and designate its weight by F. If we call them propellants and designate their weight by P, we should have a conflict with our designation for payload.

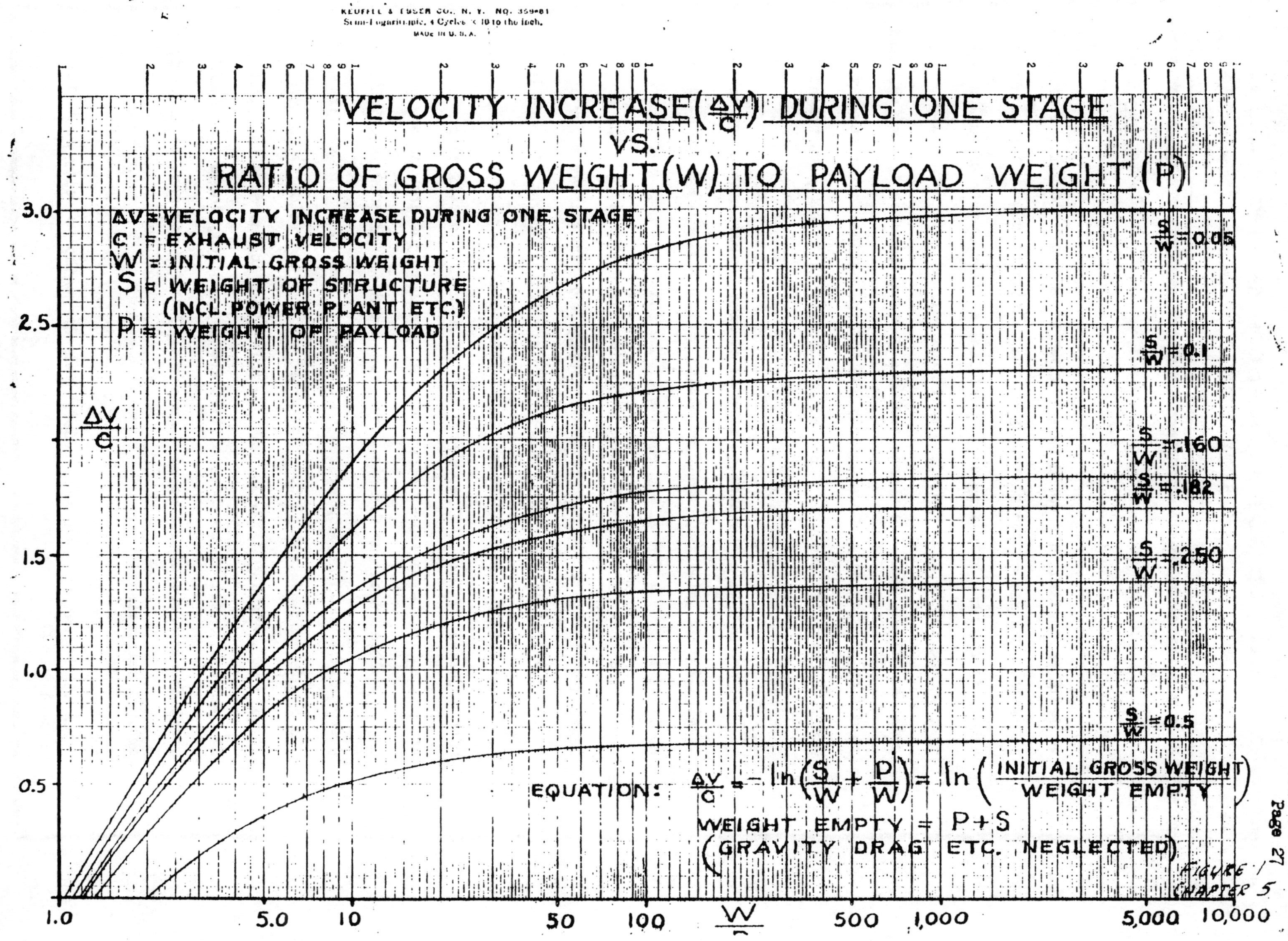

KEUFFEL & ESSER CO., N. Y. NO. 359-81
Semi-Logarithmic, 4 Cycles × 10 to the inch.
MADE IN U.S.A.
VELOCITY INCREASE ($\frac{\Delta V}{C}$) DURING ONE STAGE
VS.
RATIO OF GROSS WEIGHT (W) TO PAYLOAD WEIGHT (P)
ΔV = VELOCITY INCREASE DURING ONE STAGE
C = EXHAUST VELOCITY
W = INITIAL GROSS WEIGHT
S = WEIGHT OF STRUCTURE
(INCL. POWER PLANT ETC.)
P = WEIGHT OF PAYLOAD
$\frac{\Delta V}{C}$
3.0
2.5
1.5
1.0
0.5
$\frac{S}{W} = 0.05$
$\frac{S}{W} = 0.1$
$\frac{S}{W} = .160$
$\frac{S}{W} = .182$
$\frac{S}{W} = .250$
$\frac{S}{W} = 0.5$
EQUATION: $\frac{\Delta V}{C} = -\ln\left(\frac{S}{W} + \frac{P}{W}\right) = \ln\left(\frac{\text{INITIAL GROSS WEIGHT}}{\text{WEIGHT EMPTY}}\right)$
WEIGHT EMPTY = P + S
(GRAVITY DRAG ETC. NEGLECTED)
1.0
5.0
10
50
100
$\frac{W}{P}$
500
1,000
5,000
10,000

FIGURE 1
CHAPTER 5

ORM 25-S-1
(REV. 8-45)

PREPARED BY: P. A. Lagerstrom
F. H. Clauser   DOUGLAS AIRCRAFT COMPANY, INC.   PAGE: 28

DATE: May 2, 1946   SANTA MONICA   PLANT   MODEL: #1003

TITLE: PRELIMINARY DESIGN OF SATELLITE VEHICLE   REPORT NO. SM-11827

Chapter 5

we could accomplish this would have a gross weight of 1000 lbs., a fivefold increase. The reason for such extreme sensitivity is clear; the performance gain was made by adding a bit more fuel at the expense of payload and then enlarging the entire project until the payload returned to its original value. What was a fraction of a percent increase in fuel amounted to 80% of the payload. Consequently the multiplication factor was five. Simple clarity of reason does not alter the fact that the gross weight is a variable of questionable reliability.

We are now in a position to make an elementary examination of the feasibility of using rockets to establish new satellites. To do this we shall anticipate a few of the results of the next two chapters. There we shall find that by using alcohol and liquid oxygen (these were the propellants used in the V-2), we can obtain average exhaust velocities of about 8,500 ft./sec. and a corresponding entire structural weight of about 16% of the gross weight. Both of these figures have had a certain amount of optimism injected in them, to represent what we might reasonably expect to accomplish in the foreseeable future. If we select 500 lbs. as our goal for a payload, then our formula shows that a vehicle of 5000 lbs. initial gross weight could be accelerated to 11,420 ft. per sec. If the size of vehicle is increased to 50,000 lbs. gross weight, the velocity is 15,090 ft. per sec. and a 500,000 lb. vehicle only gives an increase to 15,510 ft. per sec.* All of these velocities are impressively large, but fall considerably short of our round figure of 24,500 ft. per sec. required for orbital velocities.

*Even if the vehicle were made indefinitely large the velocity could not exceed 15,600 ft. per sec.

ORM 25-5-1
(REV. 8-41)
PREPARED BY: P. A. Lagerstrom
F. H. Clauser
DATE: May 2, 1946
TITLE: PRELIMINARY DESIGN OF SATELLITE VEHICLE

DOUGLAS AIRCRAFT COMPANY, INC.
SANTA MONICA PLANT

PAGE: 29
MODEL: #1003
REPORT NO. SM-11827

The question immediately arises: By using liquid hydrogen, the fuel that tops the list with an exhaust velocity of about 13,500 ft./sec., can we achieve our desired velocity? Unfortunately, liquid hydrogen has a number of characteristics (which will be discussed in detail later) that necessarily cause an increase in structural weight. Our figure of 16% for structural weight is increased to 25% for use with liquid hydrogen. The following table summarizes the velocities calculated for both alcohol and liquid hydrogen:

Gross wt. for 500 lb. payload	Velocity of Vehicle Using Alcohol	Velocity of Vehicle Using Liquid Hydrogen
5,000 lbs.	11,420 ft./sec.	14,180 ft./sec.
50,000 lbs.	15,090 ft./sec.	18,160 ft./sec.
500,000 lbs.	15,510 ft./sec.	18,620 ft./sec.
Indefinitely large	15,600 ft./sec.	18,700 ft./sec.

The liquid hydrogen shows improvement[*] over the alcohol but is still considerably short of producing the orbital velocity figure of 24,500 ft. per sec.

We are forced to conclude that a realistic appraisal of the problem shows that our technology, even allowing a reasonable note of optimism to creep in, has not sufficiently advanced as yet to permit us to build a single unassisted vehicle capable of acquiring sufficient speed to remain in space as a satellite. This is doubly emphasized when we remember that as yet we have neglected entirely the effects of air resistance and gravity.

Since we cannot attain our goal with an unassisted vehicle, we next examine the problem of giving the vehicle enough initial speed so that it

[*]This is true only for single assisted vehicles using the simplified analysis presented here. When the multistage vehicles (to be considered presently) are compared using a refined analysis, the conclusion is different.

ORM 25-5-1
(REV. 8-43)
PREPARED BY: P. A. Lagerstrom
F. H. Clauser

DOUGLAS AIRCRAFT COMPANY, INC.

PAGE: 30

DATE: May 2, 1946   SANTA MONICA   PLANT

MODEL: #1003

TITLE: PRELIMINARY DESIGN OF SATELLITE VEHICLE

REPORT NO. #SM-11827

Chapter 5

can subsequently attain orbital velocities under its own power. Since rocket power plants have been shown capable of supplying more than half of the velocity required, it appears logical to ask if they cannot be used to supply the other half. To answer this question in the affirmative we introduce the concept of a multistage rocket. We shall find this idea fundamental to our later work. To illustrate this concept, let us consider a two-stage rocket. The primary vehicle will be carried along as the "payload" of a larger secondary vehicle. When this larger vehicle has exhausted its fuel, and hence its usefulness, it will be discarded and the smaller vehicle will continue to accelerate under its own power, adding its own velocity increase to that imparted by the larger stage. The particular example selected above is a special case of a much more general idea, namely, that of discarding weight once it has served its purpose and is no longer necessary. A moment's reflection shows that this can be of great aid, because as the fuel is used up, the structural weight and the payload, initially insignificant, become major items and if substantial reductions in the structural weight are possible at this point, the remaining fuel will be capable of supplying correspondingly greater accelerations and velocities.

In place of the method proposed above, it is conceivable that the fuel could be contained in multiple tanks and as each is drained in turn, it and its associated structure would be jettisoned. With this reduced weight, the acceleration would increase considerably and it might be desirable to shut down a portion of the rocket power plant to keep the loads on the remainder of the structure within reasonable limits. The remainder

Chapter 5

of the fuel would be used to produce smaller thrust over a longer period of time. If this is done, it is of course advisable also to jettison the idle power plants.*

It will be readily appreciated that such staging schemes are limited only by the fertility and ingenuity of the designer's imagination. For the sake of definiteness, we have confined our attention in this report to the clearcut scheme originally proposed, but it is not intended to imply that this is a final arrangement.

Let us return to the problem of examining the possibilities of achieving orbital velocities. We found for a single stage, that $V = -c \ln \frac{S + P}{W}$. If we now have a two stage rocket, and we designate the larger vehicle, which is fired first, by the subscript 1 and the smaller by the subscript 2, then

$$V_{total} = -c \ln \frac{S_1 + W_2}{W_1} - c \ln \frac{S_2 + P_2}{W_2} .$$

Here we have used W_2, the gross weight of the smaller vehicle, as the payload of the larger. With this notation W_1 is the gross weight of the entire aggregate.

It is now logical to ask: For a given payload and a fixed value of aggregate weight, what is the correct proportioning of the two stages to give the greatest total velocity? If the large stage can be built so that its entire structural weight is the same percentage of its gross weight as that of the smaller stage, then simple differentiation shows

---

*Viewed from this standpoint, our original proposal of a series of progressively larger vehicles each carrying the preceding member as payload, consists of building tanks, power plants and structure in associated size units and jettisoning them as units.

Chapter 5

the greatest total velocity is obtained when the ratio of payload to gross weight is the same for each stage, i.e. $\frac{W_2}{W_1} = \frac{P}{W_2}$.

If we apply these results to our alcohol and liquid oxygen powered vehicle, (and assume that the entire structure of the large stage can also be built for 16% of its gross weight) we can achieve the following velocities with the corresponding aggregate combinations:

TWO STAGE ROCKET VEHICLE, USING ALCOHOL AND LIQUID OXYGEN AND

CARRYING A PAYLOAD OF 500 LBS.

Gross Weight of 1st Stage	Gross Weight of 2nd Stage	Total Velocity
50,000 lbs.	5,000 lbs.	22,849 ft./sec.
5,000,000 lbs.	50,000 lbs.	30,180 ft./sec.
500,000,000 lbs.	500,000 lbs.	31,020 ft./sec.

This table illustrates two salient points:

1st, a two stage rocket vehicle, using feasible values of exhaust velocity and structural weights has been shown to have a reasonable margin over the minimum essential requirement to attain orbital speeds. It only remains to be seen if this margin is sufficient to account for the effects of air resistance, gravity and the like.

2nd, we notice, upon comparing this table with the results of our single stage calculations that for a given total weight, (e.g. 50,000 lbs.) we can attain a greater total velocity from two stages (22,840 ft./sec.) than we can from one stage (15,090 ft./sec.). And this is in spite of the fact that we have the weight of two machines instead of one.

Chapter 5

This second point immediately poses the following questions: Is it always better to use two stages than one? If two stages are superior, would three or more stages give even greater velocities for a fixed aggregate weight? These questions are answered by the accompanying graph,* on which has been plotted the total velocity with the available exhaust velocity taken as a unit, against the gross weight of the aggregate for a one pound payload for 1, 2, 3, 4, and 5 stages. This has been computed on the assumption that each vehicle could be built for an entire structural weight of 16% of its gross weight. In each case, the stages have the optimum proportions mentioned above.

We see immediately that two stages are not always superior to one. For small aggregate weights, a single stage is better, but at higher weights the two-stage curve crosses over and gives higher velocities. For a better understanding of the reasons behind this it is helpful to refer back to our remarks on the great variability of the gross weight of a single stage. There we saw that in our attempt to get higher and higher performances from a fixed exhaust velocity, we were exchanging payload for fuel and then swelling the size of the entire vehicle to return the payload to its specified value. As the payload became a diminutive portion of the vehicle, its exchange for fuel could affect the performance but little, while the multiplication in size became astronomical. It is at this point of diminishing returns that it is

*For additional graphs of the same type but with $\frac{S}{W} = .1, .143, .182,$ and .25, see Chapter 8.

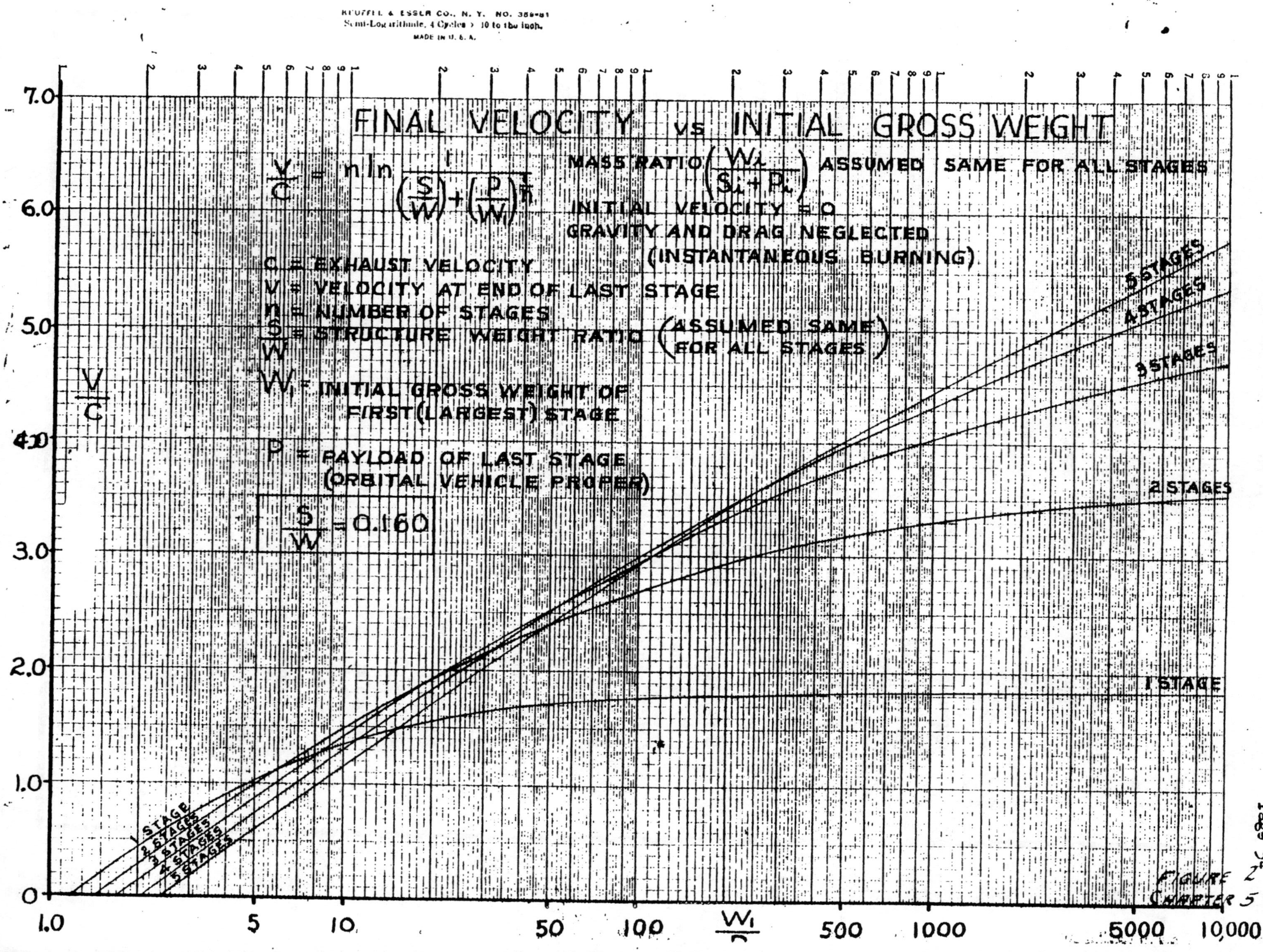
KEUFFEL & ESSER CO., N. Y. NO. 358-81
Semi-Logarithmic, 4 Cycles × 10 to the inch.
MADE IN U. S. A.
FINAL VELOCITY vs INITIAL GROSS WEIGHT
$\frac{V}{C} = n \ln \frac{1}{\left(\frac{S}{W}\right) + \left(\frac{P}{W_1}\right)^{\frac{1}{n}}}$
MASS RATIO $\left(\frac{W_i}{S_i + P_i}\right)$ ASSUMED SAME FOR ALL STAGES
INITIAL VELOCITY = O
GRAVITY AND DRAG NEGLECTED
(INSTANTANEOUS BURNING)
C = EXHAUST VELOCITY
V = VELOCITY AT END OF LAST STAGE
n = NUMBER OF STAGES
$\frac{S}{W}$ = STRUCTURE WEIGHT RATIO (ASSUMED SAME FOR ALL STAGES)
W_1 = INITIAL GROSS WEIGHT OF FIRST (LARGEST) STAGE
P = PAYLOAD OF LAST STAGE (ORBITAL VEHICLE PROPER)
$\frac{S}{W} = 0.160$
$\frac{V}{C}$
$\frac{W_1}{P}$
1 STAGE
2 STAGES
3 STAGES
4 STAGES
5 STAGES
0
1.0
2.0
3.0
4.0
5.0
6.0
7.0
1.0
5
10
50
100
500
1000
5000
10,000

FIGURE 2
CHAPTER 5

Chapter 5

better to use two stages. This same line of reasoning answers our question about larger numbers of stages, because as the two-stage vehicle reaches its point of diminishing returns, it is advantageous to use 3 stages and so on for 4, 5, 6 and higher numbers of stages. It is interesting to note that this simplified analysis would indicate that the Germans could have accomplished the mission of the V-2's with an approximate 25% decrease in total weight if they had used two stages instead of one. Undoubtedly, with all factors taken into account, including the urgency of the situation, they were well justified in using a single stage missile.

Thus far, by neglecting the "practical" details of gravity, air resistance, variation of exhaust velocity with altitude, inclination of flight path, control, maneuvering and the like, we have indicated the possibility that our technology has advanced sufficiently for us to launch a new satellite into space. Now we must determine how great will be the influence of these "practical" details.

First, let us consider the effect of gravity. So far, it has made no difference whether we used our fuel to produce a large thrust for a short time or a small thrust for a longer time. All that mattered was the velocity of the exhaust products and not the consumption rate. However, when the vehicle is accelerating vertically upward, this is no longer the case. If the thrust is insufficiently large to exceed the weight of the vehicle, the rocket will ineffectually expel its fuel, accomplishing little more than a display of fireworks. It can easily be shown that for vertical acceleration, larger

Chapter 5

velocities will be attained as greater thrusts are used for shorter times. As the thrust becomes infinite, the velocity will approach that calculated by our simplified analysis. This concept of an infinite thrust, frequently encountered in more abstract treatises on rocket vehicles, would indicate that the effect of gravity could be made negligibly small. However, closer examination shows this is not the case. As we increase the thrust, the weight of the rocket combustion chamber, pumps, piping, controls and associated structure goes up. Furthermore, the remaining structure such as tanks and supports is subjected to increasing loads as the thrust increases, with a consequent increase in weight of these items. Since we have seen that the performance is critically sensitive to the structural weight ratio S/W, the increase of this parameter will rapidly nullify the benefits of increased acceleration; in fact, we would anticipate that an optimum acceleration exists, representing the best compromise between the advantages of high thrust and the accompanying disadvantages of high structural weight. Unfortunately we have not as yet laid the foundation of structural analysis necessary to pursue this investigation further at this point.

If we attempt to examine the other "practical" factors in detail we shall find that corresponding foundation data are lacking for them too. Consequently, it is advisable to turn our attention now to a detailed examination of the capabilities of the rocket power plant and an analysis of the feasible weights of structures. Later we shall resume our investigation of the "practical" details. In the analytical work that preceded the writing of this report, performance studies,

Chapter 5

structural analysis, and the assessment of rocket power plant capabilities all proceeded hand in hand. Consequently, in the next two chapters, which deal in turn with rockets and structural weights, we shall find frequent references to the results of our more detailed performance analysis which will be presented later. Unfortunately there appears to be no way of avoiding this lack of straightforwardness in the presentation of a subject whose parts are so closely interrelated.

As an aid to the reader, a few words of coordination may prove helpful. It was decided to investigate two vehicles. One employed alcohol and liquid oxygen rockets as representative of an established technique founded on the Germans' experience with the V-2. The second employed liquid hydrogen and liquid oxygen rockets as representative of the top class of high velocity propellants. It was found best to use a four stage vehicle when using alcohol and oxygen and a two stage vehicle when using hydrogen and oxygen.

Chapter 6

6. ROCKET POWER PLANTS AND FUELS

The importance of selecting propellants which give high exhaust velocities is obvious from Chapter 5. High exhaust velocity cannot be the sole criterion, however. One or both propellants of every system proposed to date possesses physical properties which are so extreme as to present major engineering or operational problems, in some cases, to a degree almost precluding use of the propellant. Consequently, along with a consideration of specific impulse must go a careful weighing of the other advantages and disadvantages of a particular system. The disadvantages of some properties such as inflammability, corrosivity, toxicity, sensitivity to detonation, availability and handling and storing qualities are obvious. Others, such as high vapor pressure, low density, low boiling point, high average molecular weight of the products of combustion are not so obvious and require a few words of explanation.

Two types of liquid propellant systems are used: bipropellant and monopropellant. In the bipropellant system a fuel and an oxidizer, both of which may be a mixture of two or more compounds, are mixed and burned in the combustion chamber. In the case of the monopropellant system a liquid or a mixture, which is stable at ordinary temperatures, is injected into the combustion chamber where, after ignition it decomposes at the temperatures and pressures prevailing. The bipropellant system is more complicated than the monopropellant since it presents problems of designing injectors to give good mixing, of feeding the propellants at a constant mixture ratio and of providing tanks, tubing and pumps for two propellants. The monopropellants have, in general, lower specific im-

FORM 25-5-1
(REV. 8-43)

PREPARED BY: G. H. Peebles   DOUGLAS AIRCRAFT COMPANY, INC.   PAGE: 39

DATE: May 2, 1946   SANTA MONICA   PLANT   MODEL: #1033

TITLE: PRELIMINARY DESIGN OF SATELLITE VEHICLE   REPORT NO. SM-11827

Chapter 6

pulses and are inherently unstable, decomposing explosively under high temperatures or shock.

Motors operating continuously for periods longer than about 30 seconds must be provided with cooling. One method, known as regenerative cooling, brings one or both propellants to the combustion chamber through ducts in the motor walls. This system is limited by the ability of the propellants to absorb the necessary heat without boiling or decomposing. Another method called film cooling injects a liquid, preferably one of the propellants, through numerous small orifices so as to provide a cool film between the hot gases and the motor walls. This system was used on the German V-2 motor in addition to regenerative cooling with alcohol. Temperature of the gases may also be reduced by using an excess of fuel or oxidizer, by addition of water to the fuel, or by injecting water directly into the chamber. If carried to extreme the latter methods are costly in specific impulse.

Unless gas pressurization is used pumps are required to supply the propellants to the combustion chamber at high pressures and mass flow rates. To keep the weight of the pumping system low it is desirable to use high speed centrifugal pumps and as few pump stages as possible. Weight saving along these lines is limited by cavitation. Since cavitation appears on the blade at the point where the pressure drops to the vapor pressure of the fluid, a propellant with high vapor pressure leads to lower rotative speeds and more stages and so to excessive feed system weights. Low density of the propellant also increases pump weights. This is due to the fact that lower density reduces the pressure rise

Chapter 6

per stage of a centrifugal, pump so that more stages are required.

The specific impulse of a propellant system at optimum expansion ratio can be calculated from the formula

$$I = 6.94 \sqrt{\frac{T_c}{M}} \left\{ \frac{2\gamma}{\gamma - 1} \left[1 - \left(\frac{p_e}{p_c} \right)^{\frac{\gamma - 1}{\gamma}} \right] \right\}^{\frac{1}{2}} ,$$

if T_c, M and γ (respectively the temperature, average molecular weight, and ratio of the specific heats of the gases in the chamber) are known for the pressure ratio p_e/p_c. Now p_e/p_c, with minor reservations, can be chosen without regard to the propellant system and, although γ varies some for the different systems, its effect is comparatively small, Hence the ratio T_c/M accounts for the major part of the variation in specific impulse exhibited in table (1). Stoichiometric mixture ratios give a maximum for T_c but not necessarily for T_c/M. For example, stoichiometric mixture ratio for liquid hydrogen and liquid oxygen occurs at approximately 89% by weight of oxygen, but figure (1) shows that the maximum of T_c/M as reflected in I lies at about 76% which corresponds to nearly five moles of hydrogen to one of oxygen instead of the stoichiometric ratio of 2 to 1. The reason, of course, is that the low molecular weight of the excess hydrogen in the gases reduces M and more than offsets the decrease in temperature.

Chapter 6

TABLE 1. - Summary of Rocket Propellants[*]

Bipropellant Systems

System (wt. percent)	Spec. Grav.	P_c atmos.	T_c °R.	T_e °R.	M	I sec.
(1) 23.9% liquid hydrogen, 76.1 liquid oxygen	.248	23.0	4,960	2,650	8.36	362
(2) 45.8% hydrazine, 54.2% liquid fluorine	1.061	20.4	6,850			292
(3) 60.1% hydrazine, 39.9% liquid oxygen	1.061	20.4	5,550	3,090		264.0
(4) 32.6% methyl amine, 67.4% liquid oxygen	.985	20.4	6,100	3,560		251.5
(5) 31.9% liquid ammonia, 8.1% liquid acetyline, 60% liquid oxygen		20.4	5,880			257
(6) 25.4% liquid acetylene, 74.6% liquid nitrogen tetraoxide		23.0	6,230	3,960		256
(7) 41.5% liquid ammonia, 61.6% liquid oxygen		20.4				249
(8) 58.5% hydrazine, 58.6% hydrogen peroxide	1.237	20.4	4,890	2,900		249
(9) 9.8% liquid acetylene, 21.1% liquid ammonia, 69.1% liquid nitrogen tetraoxide		23	5,530	3,600		244
(10) 40% ethyl alcohol, 60% liquid oxygen	.966	20.4	5,720			243
(11) 71.5% liquid oxygen, 28.5% gasoline	.978	20.4	5,930	3,460	22.66	242.0
(12) 24.0% liquid acetylene, 31.4% liquid ammonia, 44.6% liquid oxygen		21.4	4,140	2,070		240
(13) 19.4% liquid propane, 80.6% liquid nitrogen tetraoxide		23	5,580	3,600		238

[*]Table I (slightly revised) from "Fuel Systems for Jet Propulsion" (prepared for Commander in Chief, U. S. Fleet) by Alexis W. Lemmon. Jr.

Chapter 6

TABLE 1. - Summary of Rocket Propellants (Cont.)

Bipropellant Systems

	System (wt. percent)	Spec. Grav.	P_c atmos.	T_c °R.	T_e °R.	M	I sec.
(14)	46.6% liquid ethylene, 53.4% liquid oxygen	.774	20.4	4,040		15.00	236
(15)	40% nitromethane, 60% hydrogen peroxide		20.4	5,350		24.3	227
(16)	70% nitromethane, 21% hydrogen peroxide, 4% water, 5% methyl alcohol		20.4	4,950		21.1	226
(17)	92.9% nitromethane, 7.1% liquid oxygen	1.139	20.4	5,160	2,910		225.5
(18)	21.44% methyl alcohol, 78.56% hydrogen peroxide	1.239	20.4	4,590	2,960		225
(19)	22.2% gasoline, 54.5% liquid oxygen, 23.3% liquid nitrogen	.931	20.4	5,290	3,020	23.92	221.5
(20)	57.1% methyl alcohol 42.9% liquid oxygen	.911	20.4	4,120	2,350		221
(21)	25% aniline, 75% red fuming nitric acid	1.390	20.4	5,525		25.41	220.5
(22)	17.9% mono-ethyl aniline, 82.1% mixed acid	1.396	23.0	5,060	3,400		210.0
(23)	33.6% liquid diborane, 66.4% water	.706	20.4				200
(24)	24.4% ethylene diamine, 55.4% hydrogen peroxide, 20.2% water	1.174	20.4	3,140	1,780		196.3
(25)	48.4% liquid ethane, 51.6% liquid oxygen	.760	20.4	1,910		12.40	180

Chapter 6

TABLE 1. - Summary of Rocket Propellants (Cont.)

Monopropellant Systems

System (wt. percent)	Spec. Grav.	P_c atmos.	T_c °R.	T_e °R.	M	I sec.
(1) Nitromethane (100%)	1.139	20.4	4,590		20.3	222
(2) 80% methyl nitrate, 20% methyl alcohol		20.4	4,370		20.0	221
(3) 70% nitroglycerine, 30% nitrobenzene		20.4	4,950		22.9	217
(4) Nitromethane (100%)	1.139	20.4	4,430	2,400	20.34	216.5
(5) Diethylene-glycol dinitrate (100%)	1.483	20.4	4,590		21.8	215
(6) Diethylene-glycol dinitrate (100%)	1.483	20.4	4,540	2,520		213.1
(7) 89.6% nitromethane, 10.4% nitrobenzene	1.181	20.4	4,450	2,400		212
(8) 90% nitromethane, 10% nitrobenzene		20.4	3,980		19.4	211
(9) 83% nitromethane, 17% nitroethane	1.123	20.4	3,940	2,050		206
(10) 90% diethylene-glycol dinitrate, 10% nitro-benzene		20.4	3,960		20.6	204
(11) Ethyl nitrate (100%)		20.4	3,530		18.2	203
(12) 61.9% nitromethane, 38.1% nitroethane	1.105	20.4	3,310	1,650		195.9
(13) Hydrogen peroxide (100%)	1.463	20.4	2,258	1,173	22.68	146.3
(14) 87% hydrogen peroxide, 13% water	1.381	20.4	1,668	840		126.3

Analysis PRELIMINARY DESIGN OF SATELLITE VEHICLE

DOUGLAS AIRCRAFT COMPANY, INC.

Prepared by G. H. Peebles

Date May 2, 1946

Page 44

Model #1033

Plant

Report No.

PERFORMANCE CHARACTERISTICS OF THE LIQUID
HYDROGEN — OXYGEN SYSTEM AT 23 ATMOSPHERES

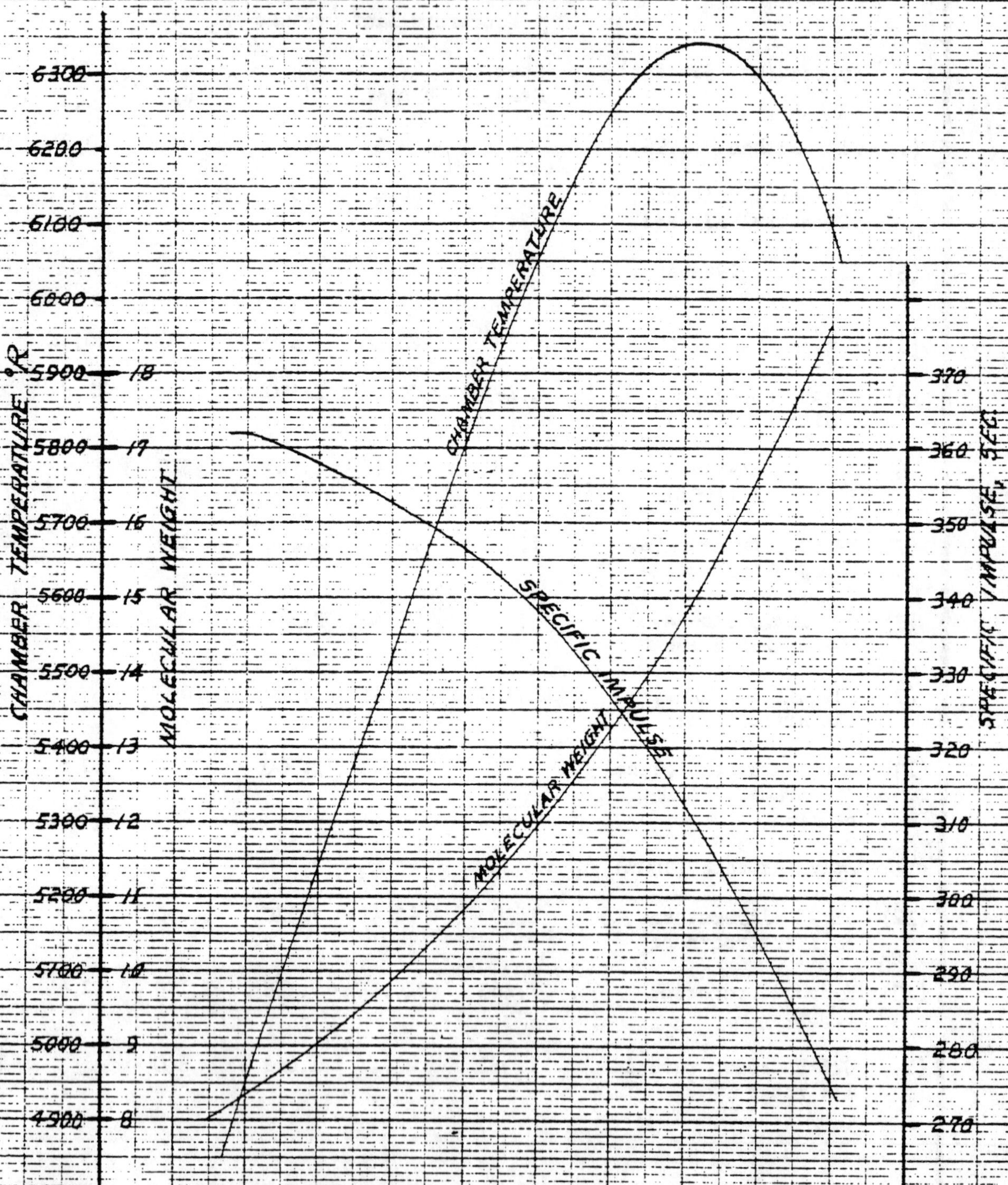

WEIGHT PERCENT LIQUID OXYGEN

FIGURE 1

CHAPTER 6

Chapter 6

The quantities T_c, M and γ of formula (1) are calculable for a given chamber pressure. Their values along with specific impulse and density are given in table (1) for a number of propellant systems. This list does not contain all possible systems but is representative of rockets obtaining their energy from combustion. At first glance it might seem that, in view of the variety of fuels available for consideration, the performance might well rise beyond the limits indicated by the table as unnoticed fuels with higher heats of combustion are brought to attention. That the problem is not quite so simple is shown by comparison of liquid oxygen-alcohol with liquid oxygen-gasoline. The heat of combustion of gasoline is appreciably higher (60% higher if n-octane is used for gasoline) than ethyl alcohol. Yet the specific impulses of the two systems are approximately equal. The underlying reason is the appearance of dissociation at about 4500°R which absorbs large amounts of energy. Both systems are composed of the elements carbon, hydrogen and oxygen. In addition to the equations of oxidization of alcohol and gasoline to carbon dioxide and water, the reversible reactions of dissociation

$$2\ H_2O \rightleftharpoons H_2 + 2OH,$$
$$H_2O \rightleftharpoons H_2 + O\ ,$$
$$2\ H_2O \rightleftharpoons 2H_2 + O_2\ ,$$
$$H_2 \rightleftharpoons 2H\ ,$$
$$CO_2 + H_2 \rightleftharpoons CO + H_2O,$$

enter into the equilibrium of the products of combustion for the two cases. The dissociation processes are accompanied by the absorption of large amounts of heat so that the greater heat of combustion of gasoline is absorbed chiefly by increased dissociation. From this example of a

Chapter 6

general behaviour it is evident that the common oxidizers and fuels composed principally of carbon, hydrogen, oxygen, and nitrogen must be fairly represented by the examples of table (1) since the same dissociations must appear to limit the chamber temperatures. The use of liquid ozone, for instance, which has a negative heat of formation, instead of liquid oxygen may increase specific impulses but not significantly.

Since common fuels and oxidizers promise nothing phenomenal it is natural to examine the uncommon reactants. If we turn to the halogens, fluorine is the logical choice because of its low molecular weight. Furthermore, hydrogen fluoride dissociates less easily than water. High molecular weights, however, limit the choice of fuels to the non-carbonaceous, since the best carbon compound which could appear in the products of combustions is carbon tetrafluoride which has a molecular weight of 88, twice that of its oxygen-formed counterpart, carbon dioxide. Another disadvantage of fluorine is that it is one of the most reactive substances known and therefore extremely difficult to handle and store. Also hydrogen fluoride is sufficiently toxic to have had consideration as a weapon of chemical warfare.

Chapter 6

Metals have also been considered as fuels because of their high heats of combustion. However when the molecular weights of their oxides are weighted against their heats of combustion it is not clear that this approach leads to higher specific impulses. Dr. A. J. Stosick* calls attention to the fact that the heat of formation of the gaseous form of the metal oxides is considerably less than that of the solid form. The latter is the one usually quoted in the present connection.

Table (1) shows that the specific impulse of liquid hydrogen and liquid oxygen exceeds by an appreciable margin that of any other system listed. If we consider that liquid oxygen is pure oxidizer, that T_c cannot be increased appreciably and that an excess of hydrogen is the most effective practical means of obtaining low average molecular weight, it seems probable that the oxygen - hydrogen system will maintain its theoretical supremacy in specific impulse for some time to come. The system has, however, a large number of disadvantages which must be overcome before use in a rocket motor. To begin with the density of the system is far below that of any other system. Low density increases the size and therefore the weight and drag of the vehicle. The boiling point of hydrogen is of course very low, $-259.18°C$. which means that the vapor pressure will be high. The combination of high vapor pressure and low density makes a light weight simple pumping system almost impossible.

*Most of the above argument against the likelihood of remarkable propellant systems was gained through a verbal discussion with Dr. Stosick of GALCIT Jet Propulsion Laboratory. Any inaccuracies in fact or theory must be charged to misunderstanding or misquotation on the part of the writer.

Chapter 6

The low boiling point, small temperature range (6.4°C) of the liquid phase and low heat of vaporization of liquid hydrogen make almost imperative use of thermos piping and pumps. High diffusivity of hydrogen makes sealing leads nearly impossible which, combined with the fact that hydrogen and oxygen are violently explosive in mixture ratios anywhere from 2% to 98%, makes accidents inevitably frequent. Cooling a hydrogen rocket is especially difficult. Neither liquid oxygen nor liquid hydrogen are usable for regenerative cooling because of their low boiling points. Liquid oxygen could not be used for film cooling since the more nearly stoichiometric mixture formed along the wall from the excess hydrogen would give intense heat. Nor could liquid hydrogen be used since it diffuses too rapidly to form an insulating film. Experimental investigation is difficult and hazardous because the excess hydrogen in the rocket exhaust forms a huge ball of flame on coming into contact with the atmosphere.

The difficulties enumerated tend to reduce the effective engineering use of the hydrogen-oxygen motor. In fact, the German V-2 engineers, from whom some of the information on hydrogen and oxygen was obtained, state that comparative designs, made for a rocket using hydrogen-oxygen and for the final rocket using alcohol-oxygen showed that the alcohol-oxygen rocket was superior in overall performance when all factors were taken into account.

While the difficulties of using liquid hydrogen as a fuel are discouraging, no one of them can be said to be impossible of satisfactory solution. It is conceivable that our technology may advance to a point where pumps can be replaced by a lighter pressure feed system such as a

Chapter __6__

gas generation system. The need for cooling may be reduced by more heat resistant materials and the inclusion of a third fluid for the specific purpose of film cooling, and so on down the list. Of all the disadvantages of hydrogen, only the effect of low density on the size and weight of the vehicle is an irremovable difficulty. Consequently in a study of the feasibility of a satellite vehicle, the liquid hydrogen-liquid oxygen motor must be included as an evaluation of the worth and necessity of a high performance motor.

By the same token it is desirable to include in the present study some motor with less spectacular performance, but which has had sufficient development to insure that this somewhat lower performance can actually be attained in practice. Only four systems, liquid oxygen-alcohol, acid-aniline, hydrogen peroxide-alcohol (with hydrazine hydrate) and liquid oxygen-gasoline, have passed out of the theoretical-experimental stage and become production or semi-production motors. One of these, oxygen and gasoline, is dubiously placed in this class since satisfactory cooling has not yet been achieved. The most successful motor, to date, particularly from the important standpoint of specific impulse, is the V-2 motor which used liquid oxygen and alcohol. The theoretical value of the specific impulse is seen from table (1) to be appreciably higher than any of the other four except oxygen and gasoline, the least successful of all. Consequently, if the choice of motor is restricted to those now available, both theoretical and past performance force the selection of the liquid oxygen-alcohol motor.

The theoretical value of the specific impulse of a rocket motor, as would be expected, is never reached in practice. It is generally agreed that 90% of the maximum theoretical impulse is obtainable. This figure

Chapter 6

is supported by experience with acid-aniline motor which has been subject to extensive investigation and development. By improving the cooling so as to allow use of the most favorable mixture ratio and by improving the injector system so as to obtain better mixing, the specific impulse of the acid-aniline motor has been brought up to 90% of its theoretical maximum. It is reasonable, therefore, to suppose both oxygen-alcohol and oxygen-hydrogen will ultimately be brought to the same degree of perfection which is equivalent to assuming specific impulses of about 220 and 326 sec.* respectively. In the case of oxygen-alcohol, 220sec. is not overly optimistic since the V-2 motor is estimated to have had a specific impulse of about 215 sec.

Available time permitted studies of satellite vehicles employing both an oxygen-alcohol motor with the aforementioned expected impulse of 220 sec. and an oxygen-hydrogen motor with a specific impulse of 326 sec. Had time been available a study based on the liquid oxygen-hydrazine would have been interesting as a happy mean between the high performance and excessive disadvantages of the oxygen-hydrogen motor and the lower performance but proven feasibility of the oxygen-alcohol motor. According to table (1) the specific impulse of the system with 60% hydrazine is 264 sec. and the propellant density is 1.061. Both figures are higher than those for oxygen and alcohol. The boiling point of hydrazine is 113.5°C which is also higher than alcohol. Another advantage lies in the fact that the motor can be cooled by using an excess of fuel without seriously

*These values are those obtained at sea level with a combustion chamber pressure of 20 atmospheres which is representative of current rocket designs.

Chapter 6

lowering the specific impulse; the loss in chamber temperature being partially offset by the appearance of free hydrogen in the gases. No obvious
disadvantages enter except unavailability of hydrazine in large quantities
and some toxicity. The former is said to be due to lack of commercial
demand and to be easily overcome.

A few items in connection with the propellant feed system, the chamber
pressure and the configuration of the combustion chamber and nozzle remain
for discussion. Time has not permitted a study of the feed system but,
on the basis of present knowledge and experience, turbine driven centrifugal pumps should be the most economical in weight for each stage of the
satellite vehicle. Recently the notion of using a gas generator for
providing the pressure has been considered and some work has been done on
developing the method. Such a scheme looks promising from the weight
standpoint. At present the turbine driven pump is the more advanced, although the margin is rather small since only a few pump-feed systems have
been designed.

Gases for the turbine could be generated by burning the propellants
in a firepot separate from the combustion chamber. However, gases generated from most systems must be cooled by some means such as introducing
water as a third component. A notable exception is the monopropellant
hydrogen peroxide which under the action of suitable catalysts decomposes
to steam and oxygen. This system was used to generate steam for the V-2
turbines.

An optimum chamber pressure exists for any given installation. This
optimum is fixed by two factors, the favorable increase in specific impulse
and the unfavorable increase in weight of chamber and pumping system as the

Chapter 6

chamber pressure increases. Since, for most installations, the optimum works out to be about 300 psia, this value was assumed for the satellite vehicles considered. A motor designed to operate at a given chamber pressure may also be run at a lower pressure, as long as the lower limit at which the propellants burn stably is not passed. Consequently throttling to a lower thrust is possible, a maneuver which will be seen in a later chapter to offer an advantage in reduced structural weight. Fig. (2) shows the behaviour of T, p_c and I with throttling for an acid-aniline motor. Since no suitable data were quickly available on the throttled characteristics of either alcohol-oxygen or hydrogen-oxygen rockets, we shall use for later investigations of this problem variations similar to those shown for acid-aniline.

An important parameter governing the configuration of the nozzle is the ratio of the exit area to the throat area, called the expansion ratio. For each value of p_e/p_c an expansion ratio exists for which the specific impulse is a maximum. In the case of the satellite vehicle for which the motor must operate at a constantly increasing altitude and therefore constantly decreasing pressure ratio, a compromise between the optimum expansion ratios at highest and lowest altitudes of operation of the motor must be made. Figures (3) and (4) show the variation in specific impulse with altitude, for alcohol-oxygen and hydrogen-oxygen rockets respectively. For these figures, the chamber pressures are assumed to be 20 atmospheres and the specific impulses at sea level for optimum expansion ratio are

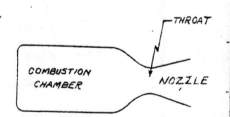

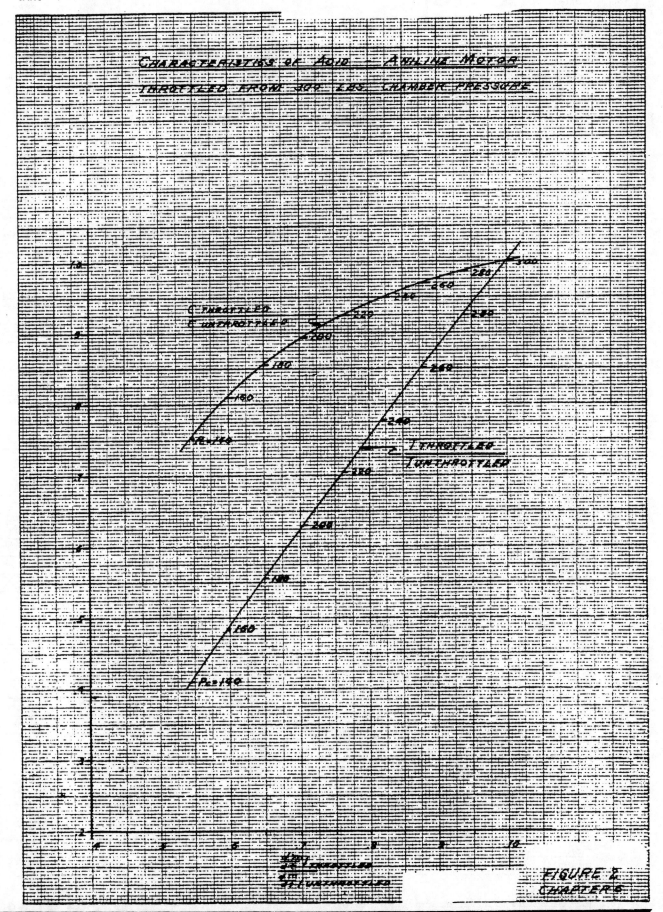

FIGURE 2
CHAPTER 6

FORM 25 BP

Analysis PRELIMINARY DESIGN OF SATELLITE VEHICLE

Prepared by G. H. Peebles   DOUGLAS AIRCRAFT COMPANY, INC.

Date   May 2, 1946

Page 5

Model #2033

Report No. SM11827

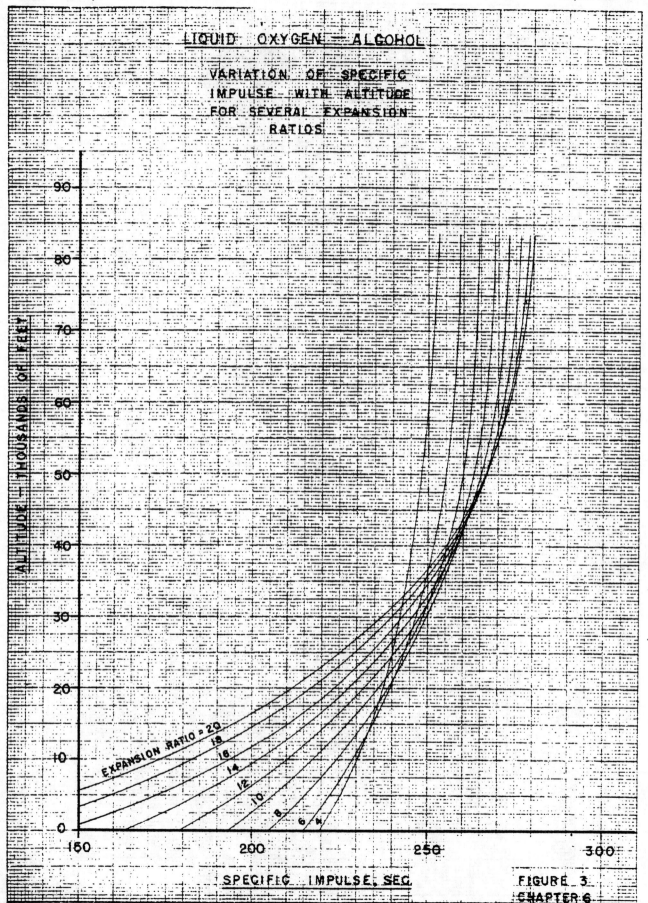

LIQUID OXYGEN — ALCOHOL

VARIATION OF SPECIFIC
IMPULSE WITH ALTITUDE
FOR SEVERAL EXPANSION
RATIOS

FIGURE 3
CHAPTER 6

ALTITUDE — THOUSANDS OF FEET

SPECIFIC IMPULSE, SEC.

EXPANSION RATIO = 20

FORM 25 8P

Analysis PRELIMINARY DESIGN OF SATELLITE VEHICLE

Prepared by G. E. Peebles   DOUGLAS AIRCRAFT COMPANY, INC.

Date   May 2, 1946

Page 55

Model #1033

Report No. SM-1627

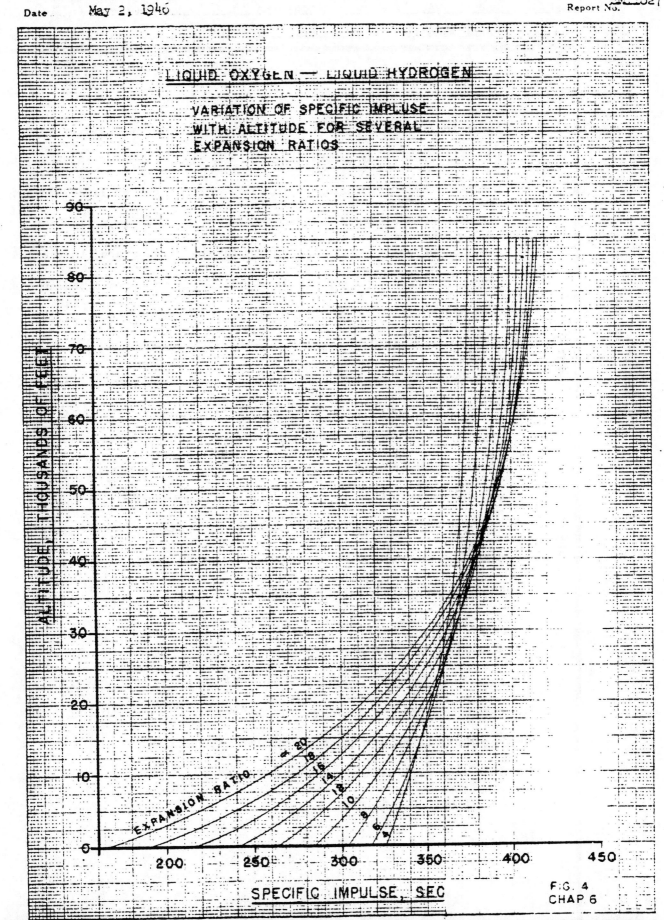

LIQUID OXYGEN — LIQUID HYDROGEN

VARIATION OF SPECIFIC IMPULSE
WITH ALTITUDE FOR SEVERAL
EXPANSION RATIOS

ALTITUDE, THOUSANDS OF FEET

EXPANSION RATIO

SPECIFIC IMPULSE, SEC

FIG. 4
CHAP 6

Chapter 6

taken as 220 and 326 seconds, in agreement with our earlier values. It is clear from these curves that for the first stage of the satellite vehicle a smaller expansion ratio is required than for later stages. In the case of the alcohol-oxygen powered vehicle an expansion ratio of 6 was used for calculating the trajectory of the first stage; in the case of hydrogen-oxygen, 8. For the later stages in both cases the expansion ratios were arbitrarily limited to 20 for the preliminary calculations. As the design progressed, it became apparent that somewhat larger expansion ratios were both desirable and possible for these later stages. However the work necessary to change the calculations at this time was felt to be unwarranted.

The chief consideration governing the shape of the combustion chamber is the necessity for allowing sufficient time for mixing and burning of the propellants while still in the chamber. As improvements in mixing are made, the dimensions of the chamber tend to decrease since the time for combustion becomes less. This principle was carried to a high degree of development by the Germans, who by the end of the war, succeeded in reducing the chamber dimensions to such an extent that they were able to use the so-called throatless combustion chamber shown in the sketch. For the V-2, this combustion chamber was less than half as heavy as the one used on production models.

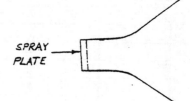

SPRAY PLATE

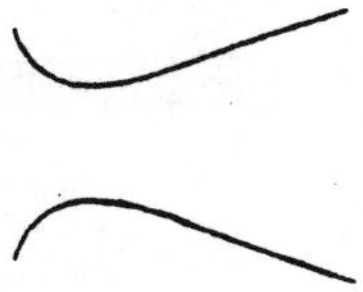

_pter 6

For most rocket motors, where the
expansion ratio is not **excessive**, it is
sufficient to use a straight conical expansion for the nozzle. However, when
expansion ratios of the order of 20 are reached, a conical diffusor must
be made undesirably long in order to avoid losses in nozzle efficiency
caused by the large radial components of
jet momentum. In order to avoid this it
is advantageous to use a nozzle shaped
as shown at the right. This type nozzle
was used on the proposed design.

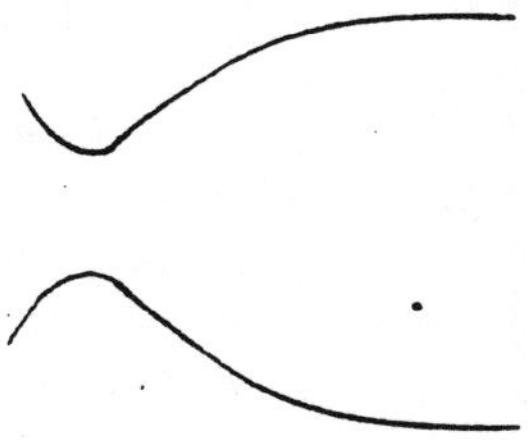

CRM 25-S-1
(REV. 8-43)
PREPARED BY: W.B. Klemperer   DOUGLAS AIRCRAFT COMPANY, INC.   PAGE:
DATE: May 2, 1946   SANTA MONICA   PLANT   MODEL: #1033
TITLE: PRELIMINARY DESIGN OF SATELLITE VEHICLE   REPORT NO. SM-11827

Chapter ___7

7. <u>CONSIDERATION OF STRUCTURAL WEIGHT</u>

<u>On The Influence of Size on Structural Weight of Rockets.</u>   If two
geometrically similar structures of different size are compared for
strength the smaller is usually the relatively stronger.   The laws governing this relationship are expressed by the doctrine of mechanical simili-
tude, which considers the dimensional correlations imposed by the invariance of certain of the physical properties involved.

Assume for instance, that the geometrical similarity extends to all
structural details, especially the degree of subdivision of structural
members.   If the loads are primarily derived from volume forces such
as weights and inertia and if the two structures will be subjected to
identical accelerations, assuming also that the structures are made of
identical materials, then the following relationships obtain between
dimensions M mass, L length, T time

M/L^3 = const   for invariant material density

L/T^2 = const   for invariant acceleration

The product of both implies

M/L^2T^2 = const, invariance of force per unit volume

Since stress is force per unit area it does not remain invariant
but increases as M/LT^2 = const x L.   Therefore where strength is governed
by stress, as for instance in members carrying tensile or bending stresses the ratio of stress to given strength increases in linear proportion with size.   To assure equal strength, the larger member will have
to be made huskier, hence heavier.   If wall thicknesses are increased,

the structural weight per volume of vehicle would tend to go up approximately in direct proportion; actually this starts a vicious circle inasmuch as the increase in gross weight will encroach on performance.

Where structural members are endangered by limits of structural stability as in column compression, there the critical carrying capacity is also increased at the rate of the square only instead of the cube of size so that the disadvantage in strength as well increases linearly with size exactly as in the case of tensile stress members. However, where strengthening can be done by increasing column diameters, this would suffice at the rate of the five-fourths power of size instead of wall thickness increase at the rate of the square of size.

Loads originating from aerodynamic action which are suffered by surface impingement, increase only proportional to the square of linear size. They can therefore, as far as velocities are invariant as integrals of acceleration ("Invariance of time scale") be suffered without additional burden. However, they will not evoke equal transverse accelerations, hence less path curvature, in inverse proportion to linear size. Hence it follows that such inertia loads as are derived from lift (and neither from thrust nor gravity) can be carried without beefing up the structural members concerned beyond proportionality with size.

Where the load components due to gravity are negligible compared to the axial inertia loads there it becomes preferable to abandon the invariance of acceleration, retaining the invariance of corresponding velocities by adopting a time scale proportional to size; $T = L$. Now

FORM 25-S-1
(REV. 8-43)

PREPARED BY: W.B. Klemperer DOUGLAS AIRCRAFT COMPANY, INC. PAGE: 60

DATE: May 2, 1946   SANTA MONICA   PLANT   MODEL: #1033

TITLE: PRELIMINARY DESIGN OF SATELLITE VEHICLE   REPORT NO. SM-11827

Chapter   7

all inertia forces will vary just like area forces; the stresses and
the strength of all structural members will be independent of size.
Thrust will also have to increase with area, not with volume, hence the
throat loading of a jet nozzle will be invariant (whereas it had to increase linearly with size under the assumption of equal acceleration).
The thrust process will now take longer in proportion to the linear size
and the range traveled under power will similarly be larger, but this
may not be detrimental. It will reduce maneuverability because the same
velocities will be attained at lesser air densities.

This analysis has a bearing on the choice of the best acceleration
peak value for a given vehicle size as this choice must be governed by
a compromise between those factors which derive an advantage from quick
acceleration and those which favor keeping it slow. The structural weight
of members carrying the inertia load belong into the latter group. In
a vehicle of the V2 (A4) type they - tanks and fuselage - are estimated
to make up about 5% of the gross weight. This weight component will
have to be expected to go up in linear proportion to axial acceleration.
The weight of the thermodynamic and mechanical machinery of the power
plant which make up about 10% of the missile's gross weight, should essentially be linearly proportional to thrust, thus similarly to axial
acceleration. Since tank loads diminish as fuel is burned at a constant
rate, their strength is dictated essentially by the initial acceleration
of their own stage or by the early surge of it occasioned by the gain
of motor thrust efficiency with altitude. However, they must also be

Chapter 7

capable of withstanding their full fuel load at the peak accelerations of all preceding stages. All subsequent stages will therefore require relatively stronger structures than the first one, unless the motors are throttled during all of the powered flight stages except the last one. This means that a margin has to be applied in any attempt to extrapolate a multi-stage aggregate from a single stage prototype.

In terms of the whole vehicle it may be desirable to strike a compromise to balance the advantages and disadvantages. A slightly heavier structure is balanced by the relatively lighter power plant when the thrust or acceleration is decreased approximately at some ratio like $L^{-2/3}$. The following table gives a rough estimate of the weight change in % of the prototype gross weight, assuming the "tanks and tank-like structures" made up 5% and the "power plant" 10% of the prototype gross weight.

Case	linear scale factor of geometrically similar enlargement:	1.4	2	2.8	4	
1	Retaining the prototype acceleration schedule, structural weight increases to: while power plant weight remains unchanged	2	5	9	15	%
2	Reducing the thrust loading inversely with enlargement, namely to: saves on power plant weight by: and incurs no increase of structural weight	70 -3	50 -5	35 -6½	25 -7½	% %
3	To offset structural weight increase by saving in power plant weight would require reduction of thrust loading to	80	64	50	40	%

However, any drastic reduction of thrust loading can only be considered where the prototype acceleration is many times gravity in the first place. It would disastrously encroach on performance when weight alone exacts a large fraction of the thrust loading. Obviously, a reduction of the apparent acceleration of the V2 to 50% of its initial value of 2 g would leave it burning itself out sitting on the ground. Actually, considerations of the influence of acceleration changes cannot be separated from performance calculations; they will be treated in considerable detail in the next chapter.

On the other hand, any complete vehicle of a typical design will be composed of various components which may be divided into several groups whose weights vary essentially with some more or less established exponents n of size, or of other characteristic parameters and these components will make up certain fractions $\alpha, \beta,$ of the gross weight.

Assume for instance, that tank weight is proportional to the nth power of the fuel weight. (It was shown above that under certain assumptions $n = 4/3$; under others part of the tank and fuselage structures may have $n = 5/4$ or some value between $4/3$ and 1; an average may well be less than 1.30). Assume that all other components weigh proportionally to gross weight. Let αW denote the fuel weight. Denote the quantities in a known prototype breakdown by index $_0$ so that the prototype fuel weight is $\alpha_0 \beta_0$ and the prototype tank weight $(1 - \alpha_0 - \beta_0) W_0$ where $\beta_0 W_0$ is the weight of everything that is neither fuel nor tank and assumed to weigh proportional to gross weight. Then in any article geometrically enlarged (except for beefing up where necessary) the fuel must be a lesser fraction of the gross weight, namely αW and the tank weight $(1 - \alpha - \beta)W$. These quantities will then compare in the proportion $(\alpha W / \alpha_0 W_0)^n$

$$\frac{(1 - \alpha - \beta)W}{(1 - \alpha_0 - \beta_0)W_0} = \left(\frac{\alpha W}{\alpha_0 W_0}\right)^n$$

hence

$$\frac{1 - \alpha - \beta}{1 - \alpha_0 - \beta_0} \cdot \left(\frac{\alpha_0}{\alpha}\right) = \left(\frac{W}{W_0}\right)^{n-1}$$

The following table of values of gross over payload for exponents from 1.3 down to 1.1 will give an idea of the order of magnitude of the reduction of fuel capacity necessary and also of the sensitivity of the result to the choice of the assumption of n.

<u>Values of gross weight/payload.</u>

n =	1.10		1.15		1.20		1.25		1.30	
β =	.20	.25	.20	.25	.20	.25	.20	.25	.20	.25
$\alpha_0 = .70$	1	1	1	1	1	1	1	1	1	1
$\alpha = .65$	130	2310	26	179	12	50	7.3	23	5.4	14
.60	5580	322000	331	4950	81	613	35	176	21	80
.55			2860	65600	415	4350	130	850	61	292
.50					1830	23500	435	3340	168	920

Actually there will also be some parts of the missile which will not require enlargement or even duplication on mother stages, for instance the "brains". These could well be taken out of the structural weights class and lumped with the ultimate payload. It is estimated that about $1\frac{1}{2}\%$ of the V2 may be in this category, which would bring a worthwhile improvement

ORM 25-S-1
(REV. 3-43)

PREPARED BY: W.B. Klemperer DOUGLAS AIRCRAFT COMPANY, INC. PAGE: 6

DATE: May 2, 1946 SANTA MONICA PLANT MODEL: #1033

TITLE: PRELIMINARY DESIGN OF SATELLITE VEHICLE REPORT NO. SM-11827

Chapter 7

of the mother stages' mass ratio and can be pitted against the weight increases entailed by enlargement previously discussed. However, the advantage thus afforded eventually fades into insignificance when enlargement is carried to extremes. The fact therefore remains that unlimited geometrical enlargement of a rocket will eventually bring a penalty in weight. This is contrary to the contention advanced by some that structural efficiency will indefinitely increase with size.

The very fact that some parts of the prototype need not be enlarged as the prototype is enlarged works a hardship when an attempt is made to reduce the size. Indeed some parts cannot be reduced proportionally or not at all. They may have attained practical or otherwise determined minimum sizes. This is a very real problem in the manufacture of miniature models. For this reason it appears that the real weight per unit volume increases towards both the small and the large end of the scale. There is an optimum somewhere in the realm of "moderate" sizes. This optimum is presumably rather flat, its exact position will sensitively depend on minor variations of the components.

Thus far only geometrically similar "blow-up" with size has been considered. The disadvantages attending this method of enlargement arise from the fact that pressures due to volume forces go up with size. This applies to all hydrostatic pressures in tanks and the critically thin supports of mass loads. It would equally apply to power plant parts built to withstand pressures, notably the burning chamber if the latter were to handle a thrust proportional to the volume through a throat

Chapter 7

area proportional only to the cross sectional area of the vehicle. The latter is not feasible thermodynamically, there being no reason why higher pressures and temperatures should become easier to handle as the article is enlarged in size. Within a limited degree it may be possible to increase the nozzle diameters more than in geometrical proportion to the rest of the vehicle, but when it surpasses the caliber then geometrical similarity of the configuration is violated.

Both the hydrostatic pressure increase and the nozzle thrust loading increase are avoided if the vehicle were to be enlarged in cross section area only and not at all in height. It would then become fatter at the rate of the square root of the gross weight increase, but all weight proportions would remain essentially the same. It is as though a plurality of the prototype vehicles were arrayed, L^2 in parallel only and not also L in series. Actually this method of fattening cannot be carried to several mother stages as the grandmother would look like a mushroom. Aerodynamic drag considerations might weigh heavily against such malformation. The idea is nevertheless fruitful in that it points the way to a compromise: As the vehicle is enlarged, it may to advantage be fattened a little, thus reducing the hydrostatic and nozzle penalties without growing out of bounds in girth. If for instance the heights (lengths) are increased by the one-fifth power and the diameter by the two-fifth, instead of each by the 1/3, the hydrostatic penalties should be reduced to 2/3, yet the fineness ratio would drop only to 1/2 for every 32 fold increase in weight. However, any such violation of geometric similarity conjures up new problems of structural subdivision, large bulkheads, anti-sloshing devices and other structural exigencies whose weight penalty has to be carefully watched lest it encroach on the gain to be derived from the whole scheme.

The question may be posed: How will a change of fuel density ρ affect the tank weight? If the increased bulk is to be accommodated by geometrically similar enlarged tanks, then the linear tank dimensions increase as $\rho^{-1/3}$; the hydrostatic pressure (at any given acceleration) will actually decrease namely as $\rho^{-1/3} \cdot \rho = \rho^{2/3}$. Since for a given material strength the wall thickness has to vary with the product of pressure by radius, and the latter varies as $\rho^{-1/3}$, the wall thickness will also decrease as $\rho^{2/3} \cdot \rho^{-1/3} = \rho^{1/3}$. The tank weight is proportional to the surface and the wall thickness viz to $\rho^{-2/3} \cdot \rho^{-1/3}$. Hence it increases with linear dimension or volume $^{1/3}$. This is a strike against liquid hydrogen fuel which weighs only 7% of hydrazine or about $8\frac{1}{2}\%$ of alcohol per unit volume. This disadvantage is aggravated by the flimsiness of the thinner walls of larger vessels.

If again the tanks are to be enlarged in width only and not in height, then the tank radius varies indirectly and the required wall thickness directly with the square root of the fuel density, so that the tank weight would remain unchanged. The progressive fattening of successive stages would rapidly grow prohibitive. On the other hand fewer stages are required to accomplish the same performance with the higher exhaust velocity of the lighter fuel and vice versa so that the overall picture may not be radically affected.

It is noteworthy that the optimum proportion of a cylindrical tank from a viewpoint of minimum wall weight to volumetric content is more squat for hydrostatic pressure than for uniform (gas) pressure. As is well known, the flat headed cylinder of least surface per volume is as high as its diameter, $(h = 2r)$.

In order to make the tank heads stand up under any uniform internal pressure, they should be bulged. Hence they would have a surface $K_s r^2 \pi$ each and a calotte volume of $K_v r^3 \pi$. For equal wall thickness the bulge radius would be twice the cylinder radius and the coefficients $K_s = 1.072$ and $K_v = .274$. The lightest shape (neglecting seams) would be attained with a cylinder height of $h = (2K_s - 3K_v)r$, here $= 1.32r$ and the total height including caps $H = 1.36r$, somewhat shallower than the flat headed cylinder.

On the other hand, if the tank is to stand hydrostatic pressure which increases linearly with height and if the walls could be suitably tapered, the weight of the top would be negligible but the bottom would have to be bulged slightly more than to the double cylinder radius if it was to be made of the same thickness as the lowest part of the cylinder wall. If we also neglect this bulge for the sake of a first rough approximation, then the weight of the tank will be indicated by

$$W = (h + r)r^2 \pi \; wnfh/s$$

where w is the specific weight of the tank material, s its allowable stress;

Chapter 7

n the load factor and f the specific weight of the fuel. Defining $H = V/r^2\pi$ by the volume V and the radius r,

$$W = (V/r^2\pi + r)\ Vwnf/s$$

$$dW/dr = (-2V/r^3\pi + 1)\ Vwnf/\pi s$$

Equating this to zero yields

$$V = \tfrac{1}{2} r^3\pi = r^2\pi h$$

$$h = \tfrac{1}{2} r$$

which is four times as squat as the square cylinder of uniform wall thickness.

. If the bottom is bulged to a radius equal to the cylinder diameter and made thicker in the sump according to the hydrostatic pressure increase, then the lightest proportion turns out to be h = .325 r and H = .593 r which is rather squat. However, the influence of a variation from the optimum is not large and other considerations such as manufacture, bracing and lid, safeguards against sloshing, etc., militate against extremely shallow vessels. The lid cannot be made weightless, the walls cannot be ideally tapered, seams and anti-sloshing means have to be provided. Hence practical vessels will probably be of proportions H/r ranging somewhere around $1 \pm \tfrac{1}{4}$.

The preceding discussion of scale effects is useful for giving an overall view of the rocket design problem. However, in order to make weight estimates for preliminary designs a greater amount of detail is necessary. An attempt is made in the following pages to consider various parts of the vehicle weight separately, applying a separate scale factor to each part. Up to the present time, the best (in fact, the only) long range rocket is the V2. For this reason the V2 is used as a basis or standard for calculation. Some features of the present multi-stage designs do not appear in V2, and separate weight allowances must be made in such cases. Since no past experience or present design practice exists for staged rockets, various reasonable appearing assumptions must be made.

It is to be expected that the art of estimating weights for long range, low acceleration rockets will progress rapidly as designs reach the layout stage on the drawing board.

Chapter   7

Weight Estimation for Rocket Design Study. As a starting point
for this study, use is made of a breakdown of weights of V-2 as given
by Gilliland*. Percentages corresponding to those weights are listed
in column (1) of Table I below. Since they correspond to a mass ratio
($\frac{\text{Gross Wt.}}{\text{(Gross Wt.-Fuel)}}$) of only 3.25, a new set of percentages has been assigned in column (2) to raise the mass ratio to 4, a ratio which was
achieved for V-2 at the end of the war.

Based on a gross weight of 27,305# as given by Gilliland, the
percentages of column (2) yield weights for the various items as given
in column (3). The "Radio and Instruments" and "Warhead" of the above
reference are lumped under "Payload" below.

Table I

V-2 Major Weight Breakdown

		(1) Mass Ratio 3.25	(2) Mass Ratio 4	(3)	
1.	Tanks and Piping	5.4%	4.5%	1230	lbs.
2.	Pumping Unit	5.7	4.7	1280	lbs.
3.	Nozzle and Combustion Chamber	3.8	2.5	680	lbs.
4.	Controls and Surfaces	4.9	4.1	1120	lbs.
5.	Fuels	69.1	75.0	20,480	lbs.
6.	"Payload"	11.1	9.2	2510	lbs.
	Total	100.0	100.0	27,305	lbs.

Although the figures of Table I above are admittedly inaccurate,
they represent the best information available to this writer at the
present time.

*Gilliland, E. R., Rocket-Powered Missiles, Jet Propelled Missiles
  Panel, May 1945, page 45.

Chapter 7

Items of Table I are further broken down in Table II, below, still following **Gilliland's outline**, but using revised figures.

Table II

V-2 "Detail" Breakdown

Item		Subdivision	Weight lbs.	Sub-totals lbs.
1. Tanks & Piping				1230
	a.	Integral Tanks	1160	
	b.	Distributing Pipes & valves	70	
2. Pumping Unit				1280
	a.	Power unit & tanks	730	
	b.	Mounting & end frame	210	
	c.	Shell structure	340	
3. Nozzle & Combustion Chamber			680	680
4. Controls & Surfaces				1120
	a.	Fins	630	
	b.	Internal controls	390	
	c.	External controls	100	
5. Fuels				20,480
	a.	Oxygen	11,500	
	b.	Alcohol	8,980	
6. "Payload"				2510
	a.	Radio compartment & frames	270	
	b.	Radio Equipment	130	
	c.	Instruments, wiring	260	
	d.	Compressed air bottles	70	
	e.	Warhead	1780	

Chapter 7

It is convenient, in the analysis which follows, to regroup the items and subdivisions of Table II, so that quantities which vary alike with scale of the vehicle can be lumped. Such a regrouping is given in Table III, with sub-totals that apply to V-2.

Table III

Weights Regrouped for Analysis

Group		Contains subdivisions (From Table II)	Weight
T.	Tanks & Structures	1a, 2b, 2c, 6a	1980
M.	Miscellaneous Structure	not in V-2	--
N.	Nozzle, chamber, pumps	1b, 2a, 3	1480
H.	Provisions for H_2	not in V-2	--
C.	Controls	4a, 4b, 4c	1120
B.	"Brains"	not in V-2	--
F.	Fuels	5a, 5b	20,480
P.	Payload	6b, 6c, 6d, 6e	2240
	Total		27,300

Groups as listed in Table III are separately discussed and analyzed below. A list of notation is given here to facilitate such discussion.

a = design acceleration for structure (no. of "g's")

f_t = applied tensile stress, #/in^2

F_t = allowable tensile stress, #/in^2

k_n = constant, may or may not be dimensionless

L = scale dimension of length

ℓ = length of fuel tank, inches

(o) = subscript, referring to V-2 as a basis for calculations

ρ = fuel density, #/in^3

Chapter 7

p = fuel pressure #/in^3

r = radius of fuel tank, inches

T, M, N, etc. = group weights as captioned in Table III

t = wall thickness of tank or shell, inches

W = Gross weight of a stage, considering the sum of all succeeding (smaller) stages as payload

Stage no.: Taken in order of firing, i.e. #1 is the largest, #4 the smallest

τ = duration of burning for a stage, seconds

Group T. Tanks and Structure

Since a high percentage of the gross weight (60 - 70% is in fuel, it is to be expected that fuel tank weights will have a major bearing on the overall structural weight. For this reason, the structural items listed by Gilliland for the V-2 projectile are all lumped and assumed to vary as the fuel tank weight.

For our purposes, the fuel tanks are assumed to be integral with the structure, rather than separate, although this point has by no means been finally settled.

Consider the fuel tanks in a manner similar to the discussion (p. 58 to 69 of this report) on the influence of size on the structural weight of rockets.

$p = k_1 \rho a l$

$f_t = F_t = \dfrac{pr}{t}$ = constant for a given wall material

$\therefore t = k_2 pr = k_3 \rho a l r$

(a) Side walls of tank

Area = $2\pi l r$

Chapter 7

$$\therefore T_a = \text{wt of side walls}$$

$$= t \times \text{area} = 2\ k_3\ \rho a\ l^2 r^2$$

(b) Tank bottom

$$\text{Area} = \pi r^2$$

$$\therefore T_b = \pi k_3\ \rho a\ \ell r^3$$

Total

$$T = T_a + T_b$$

$$= \pi k_3\ \rho a\ (2\ell^2 r^2 + \ell r^3)$$

$$F\ (= \text{fuel weight}) = \rho \pi r^2 l$$

$$\therefore \frac{T}{F} = k_3\ a(2\ell + r) = k_4 a\left(\ell + \frac{r}{2}\right)$$

From V-2:

$$T_o = 1980, \quad F_o = 20,480$$

$$\ell = 243, \quad r = 31.5 \quad a = 2.0$$

$$\therefore k_4 = .000187$$

$$\therefore \frac{T}{F} = .000187\ a\left(\ell + \frac{r}{2}\right) \qquad (1)$$

Using approximate dimensions for design studies of (a) a four stage alcohol, oxygen rocket and (b) a two stage hydrogen, oxygen vehicle, group T weights have been calculated and are given in Table IV below. In applying a value for $\underline{a}$, it should be noted that the design acceleration for the first stage is at the start of that stage (minimum acceleration, full tanks) whereas for succeeding stages the design acceleration occurs at the end of the first stage burning (maximum acceleration, full tanks).

For stage 4 of the alcohol, oxygen system, it is considered that the extrapolation from V-2 is too great for a simple scale effect formula. Therefore an independent estimate based on a reasonable minimum shell thickness is shown for this case in Table IV.

FORM 25-5-1
(REV. 8-45)

PREPARED BY: J. E. Lipp   DOUGLAS AIRCRAFT COMPANY, INC.   PAGE: 75

DATE: May 2, 1946   SANTA MONICA   PLANT   MODEL: #1033

TITLE: PRELIMINARY DESIGN OF SATELLITE VEHICLE   REPORT No. SM-11827

Chapter 7

Table IV

Group T (tank and structure weights (pounds)

Stage	4	3	2	1
Alcohol & Oxygen	120	590	4350	14,300
Hydrogen & Oxygen	- -	- -	2250	36,000

Group M.   Miscellaneous Structure

This weight group is provided to allow for structural items that do not appear in V-2. There are two major components considered. First, to allow for coupling of stages, an amount of 3% W is allowed for each stage. Second, to allow for minimum gauges and general miscellaneous, weight is assumed which is 4% W for W = 1000 pounds and zero for W = 27,000 pounds and above.

Group M weights as described above are shown in figure 1, an arbitrary curve.

Group N.   Nozzle, Chamber, Pumps

Weights placed within this group are those which depend upon the rate of fuel flow for their size. It has been found in past designs that the complete power plant varies nearly directly with the mass flow rate of fuel.

For V-2, 20,480 pounds of fuel are burned in 60 seconds and N is 1480 pounds.

$$N = 1480 = \frac{60}{20,480} \cdot \left(\frac{F}{\tau}\right)$$
$$= 4.33 \; \frac{F}{\tau} \qquad (2)$$

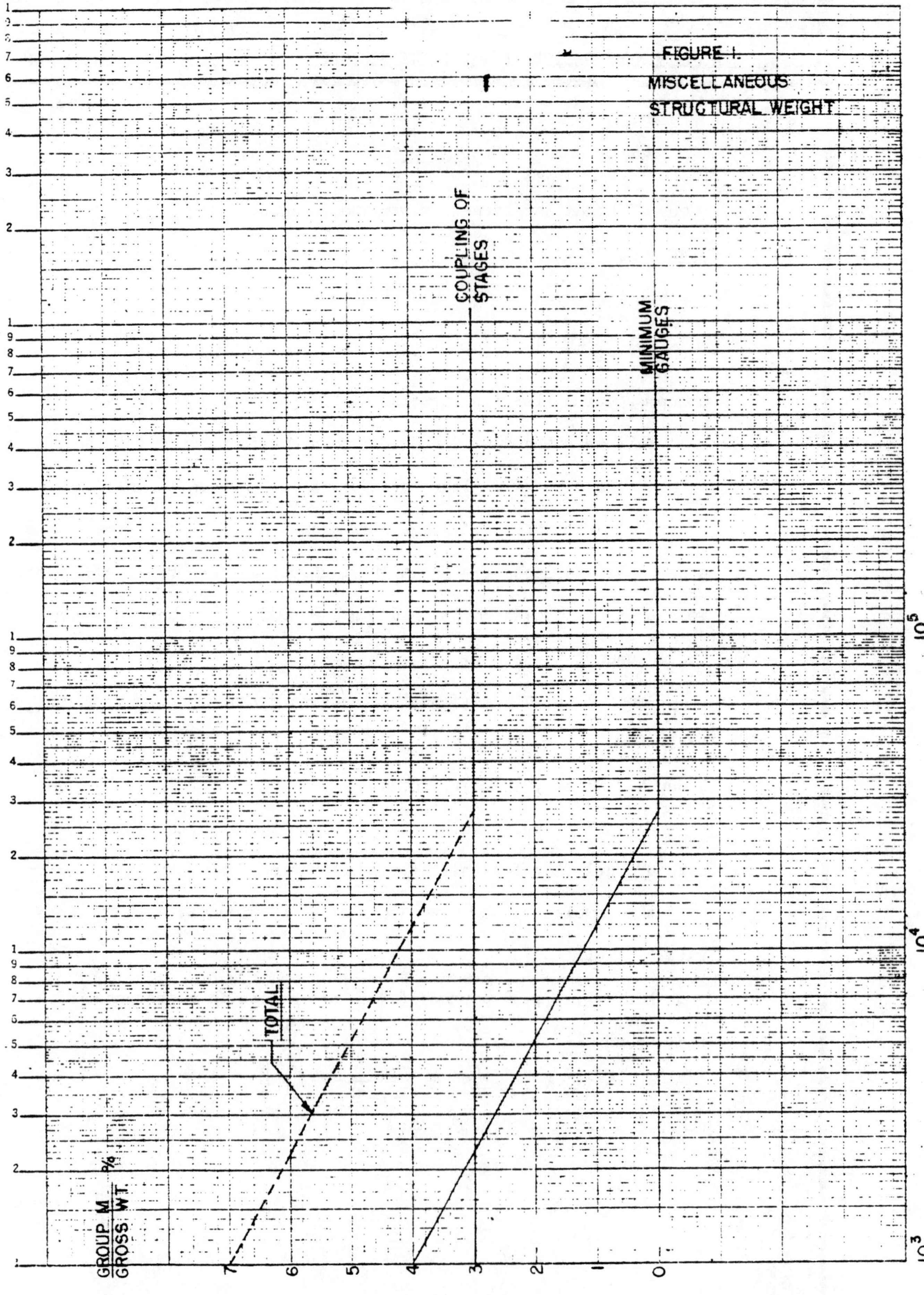

FIGURE I.

MISCELLANEOUS

STRUCTURAL WEIGHT

COUPLING OF STAGES

MINIMUM GAUGES

TOTAL

GROUP M %
GROSS WT.

Group C. Controls

Percentage of gross weight taken up by electrical, mechanical and structural controls and surfaces has been found to vary roughly with the square root of a linear scale dimension, based upon past experience with aircraft design.

$$\text{Thus} \quad \frac{C}{W} = k_5 L^{\frac{1}{2}}$$

But L varies as $W^{1/3}$

$$C = k_6 W^{7/6}$$

To avoid handling large numbers, write this in the form

$$C = C_o \left(\frac{W}{W_o}\right)^{7/6} = 1120 \left(\frac{W}{27,300}\right)^{7/6} \qquad (3)$$

Group H. Provisions for H_2

For the hydrogen burning rocket only, it is believed that special provisions will be necessary to (a) maintain the liquid state inside the hydrogen tank (b) prevent escape of the liquid and (c) prevent explosions due to the wide explosive mixture range. Although no logical basis now exists for calculating the weight of such provisions, a reasonable amount may be 1% of the gross weight of each stage.

Group B. "Brains"

By "Brains" are meant the central guiding units which furnish commands to the control system in order to guide the vehicle on its trajectory. A weight of 200 pounds is arbitrarily allowed for such equipment. This item is applied only to the last stage of each rocket system, since a single set of "Brains", with proper control system connections, should serve all stages.

FORM 25-S-1
(REV. 2-43)

PREPARED BY: J. E. Lipp   DOUGLAS AIRCRAFT COMPANY, INC.   PAGE: 78

DATE: May 2, 1946   SANTA MONICA   PLANT   MODEL: #1033

TITLE: PRELIMINARY DESIGN OF SATELLITE VEHICLE   REPORT NO. SM-11827

Chapter 7

Group F. Fuels

The total fuel weight for each stage is determined from a percentage of gross weight, which in turn is derived from trajectory calculations for the particular fuels and number of stages employed. Trajectory calculations are set forth elsewhere in this report. It will suffice to say here that for alcohol-oxygen 4-stage systems the fuel weight is taken as 60% of the gross weight for each stage whereas for hydrogen-oxygen 2-stage systems the fuel is 71% of the gross weight.

Group P. Payload

For the final stage, the payload has been set arbitrarily at 500 pounds. For other stages, the payload of each stage is the gross weight of the succeeding stages. Since the gross weight of stage 2 (say) includes the weights of stages 3 and 4 it can be said that the payload of stage 1 is simply the gross weight of stage 2, and so on for the successive stages.

Gross weight is the sum of groups T, M, N, C, H, B, F and P. Using the formulas and assumptions described above it is possible to tabulate the weights for a 4-stage alcohol-oxygen rocket and a 2-stage hydrogen-oxygen rocket. These are given in tables V and VI, respectively. Since the solution for gross weight, fuel and structure in terms of each other is a trial and error process, these figures are not completely accurate or consistent, however, they are close enough for preliminary design purposes.

Chapter 7

Table V Weight Summary, Weight in lbs.

4-stage Alcohol-Oxygen Rocket

Group	Stage	4	3	2	1
T.	Tanks & Structure	120	590	4350	14,300
M.	Miscell. Structure	160	473	1610	7,000.
N.	Nozzle, chamber, pumps	91	376	1550	7,780
C.	Controls & Surfaces	77	422	2150	11,100
H.	Provisions for H_2	None			
B.	Brains	200	- -	- -	- -
F.	Fuels	1720.	7100	32,200	140,000
P.	Payload	500	2868	11,829	53,689
	Gross	2868	11,829	53,689	233,669
	$S = T + M + N + C + H$	448	1861	9660	39,980
	S/W	.156	.157	.18	.168

Table VI Weight Summary, Weight in lbs.

2-stage Hydrogen-Oxygen Rocket

Group	Stage	2	1
T.	Tanks & Structure	2,250	36,000
M.	Miscell. Structure	555	8,740
N.	Nozzle, chamber, pumps	250	5,620
C.	Controls & Surfaces	559	16,090
H.	Provisions for H_2	150	2,750
B.	Brains	200	- -
F.	Fuels	10,900	207,000
P.	Payload	500	15,364
	Gross	15,364	291,564
	$S = T + M + N + C + H$	3764	69,200
	S/W	.246	.238

Chapter   8

8. INVESTIGATION OF DESIGN PROPORTIONS

This chapter has a double purpose. 1). The first is to continue and develop the study of the dynamics of orbital vehicles which was initiated in Chapter 5. Use will be made of the results gained in Chapters 6 and 7 concerning power plants and structural weights. 2). The second purpose is to apply the general theory thus developed to the design of two actual vehicles. In this chapter we shall be concerned only with the basic features such as number of stages, weight of stages and maximum thrusts to be used. In the following chapter, these values will be combined with the results of trajectory calculations to give a final integrated design.

General Dynamics of Orbital Vehicle. Single Stage Vehicle.- We shall improve on the analysis of Chapter 5 by taking into consideration the practical details which were left out in that chapter; namely gravity, inclination, dependency of structural weight on load factor, drag and throttling.

Gravity - First let us consider a vertical trajectory. It will be necessary to add a term, -g (the acceleration of gravity) to the right right hand side of the equation of motion, presented in Chapter 5. We obtain

$$(1) \quad \frac{dV}{dt} = - \frac{c}{m} \frac{dm}{dt} - g$$

which integrates to

$$(2) \quad V_F = c \ln \frac{m_i}{m_f} - g t_B + V_0$$

Where t_B is the burning time and subscripts "F" and "o" denote "final" and "initial" respectively.

Chapter 8

For the purposes of our later work we shall eliminate t_B in terms of the maximum load factor n defined by

(3) $\quad ng = \left(\dfrac{dV}{dt}\right) max + g$

The missile reaches its maximal acceleration (i.e. thrust - mass ratio) at the end of burning time (we assume constant thrust). Hence it follows that

$$ng = \frac{c}{S+P} \times \frac{F}{t_B} \quad or$$

(4) $\quad g t_B = \dfrac{Fc}{(S+P)n}.$

Thus (3) may be written as

(5) $\quad \dfrac{\Delta V}{c} = \ln \dfrac{W}{S+P} - \dfrac{F}{n(S+P)}$

(Note that $\dfrac{W}{S+P} = \dfrac{m_i}{m_f}$).

The following numerical example shows the importance of n. Let $\dfrac{F}{W} = .6$ and n = 6.5 (we shall later see that these are reasonable figures). Then $\ln \dfrac{W}{S+P} = \ln \dfrac{1}{1-.6} = .916$ and $\dfrac{F}{n(S+P)} = \dfrac{.6}{6.5(1-.6)} = .231 = 25\%$ of .916. Thus if the exhaust velocity c is 8,500 ft/sec, the velocity increase during one stage is 7,800 ft/sec neglecting gravity, but if gravity is considered, the velocity increase is only 75% of this or 5850 ft/sec.

<u>Inclination</u> - In most practical cases the trajectory will have variable inclination. In this case the formula for acceleration along the path of flight is

(6) $\quad \dfrac{dV}{dt} = -\dfrac{c}{m} \dfrac{dm}{dt} - g \sin \theta$

There θ is the angle the trajectory makes with the horizontal. This equation is not readily integrable unless θ is considered constant.

Chapter 8

We shall make this assumption, because in spite of its inaccuracy, it will furnish us valuable information on how θ affects our choice of n. θ will be referred to as the average inclination. Instead of (5) we then obtain

$$(7) \quad \frac{\Delta V}{c} - \ln \frac{W}{S+P} - \frac{F \sin \theta}{n(S+P)}$$

Clearly, the more horizontal the flight path, the less is the loss in performance caused by finite acceleration.

Dependency of Structural Weights on Load Factors - The example above shows that a low acceleration like 6.5g has a detrimental effect on performance. On the other hand, it was shown in Chapter 7, that a higher load factor necessitates a heavier structure and the resulting lower value of the mass-ratio parameter gives a lower value of $\ln \frac{mi}{mf}$. To study how these factors balance each other, let us consider a missile whose total initial weight W is fixed and whose weight empty may be expressed in the form

(8) $S+P = Q+R \cdot n$ where Q is that portion of the weight which is unaffected by the maximum acceleration and $R \cdot n$ is the weight of the remaining structure which is assumed to be directly proportional to n. Actually, Q represents essentially the weight of the payload and the controls, whereas $R \cdot n$ is the weight of fuel tanks, power plants and accessories, etc. Then (7) reads

$$(9) \quad \frac{\Delta V}{c} = \ln \frac{W}{Q + Rn} - \sin \theta \times \frac{W - (Q + Rn)}{n(Q + Rn)}$$

In figure 1 we see how $\frac{\Delta V}{c}$ is affected by n in a vertical trajectory where we have used the values $Q/W = .275$ and $R/W = .019$. These figures correspond to a payload which is 20% of gross weight, a structure independent of acceleration of 7.5% (.20 + .075 = .275) and a remaining structure which weighs 1.9% of gross weight for each gross acceleration.

Chapter 8

It can be seen from the figure that the advantages of high acceleration to reduce the loss from the effect of gravity are counteracted by the poorer structural weight in such a manner that an optimum acceleration exists. For this particular example, the optimum acceleration is 7.g. This could also have been obtained by the aid of calculus. Namely, if $\frac{\partial \Delta V}{\partial n}$ is equated to zero, one obtains the following cubic equation for n

$$(10) \quad \left(\frac{R}{W}\right)^2 n^3 + \frac{R}{W}\left(\frac{Q}{W} + a\frac{R}{W}\right) n^2 + 2a\frac{R}{W}\left(\frac{Q}{W} - 1\right) n +$$

$$a\frac{Q}{W}\left(\frac{Q}{W} - 1\right) = 0, \text{ where } a = \sin\theta.$$

In Figure 2 the solution of this equation has been plotted against R/W with parameters of $\frac{Q}{W}$ and θ. This chart is very useful for a rapid determination of the approximate value of the optimal acceleration. Two corrections will have to be added for a more refined analysis: 1). A correction based on a more exact weight formula. 2). A correction for drag.

From the discussion in Chapter 7 it is evident that the expression of the weight empty as a linear function of n is over-simplified. However, if we use a more accurate formula for the weight variation, we are forced to abandon general analytical methods and shall have to reduce ourselves to a numerical study of a concrete example. In choosing this example we anticipate some results to be established later. The oxygen-alcohol missile to be proposed in this report will be a four-stage missile. Its first (largest) stage will have a gross weight W_1 of 233,669 lbs., its payload (W_2) will be 11,829. We select this stage as one example. The two weights mentioned will be kept

Analysis PRELIMINARY DESIGN OF STEEL
Prepared by E.S, Rutowski   DONG   ...CRAFT COMPANY. INC.

Date   May 4, 1948   Plant

Page   84
Model   1053
Report No.   1182

FINAL VELOCITY vs. LOAD FACTOR

$$\text{EQUATION} \quad \frac{V}{C} = -\ln\left\{\frac{P+S}{W} - \frac{1 - \frac{P+S}{W}}{n\left(\frac{P+S}{W}\right)}\right\}$$

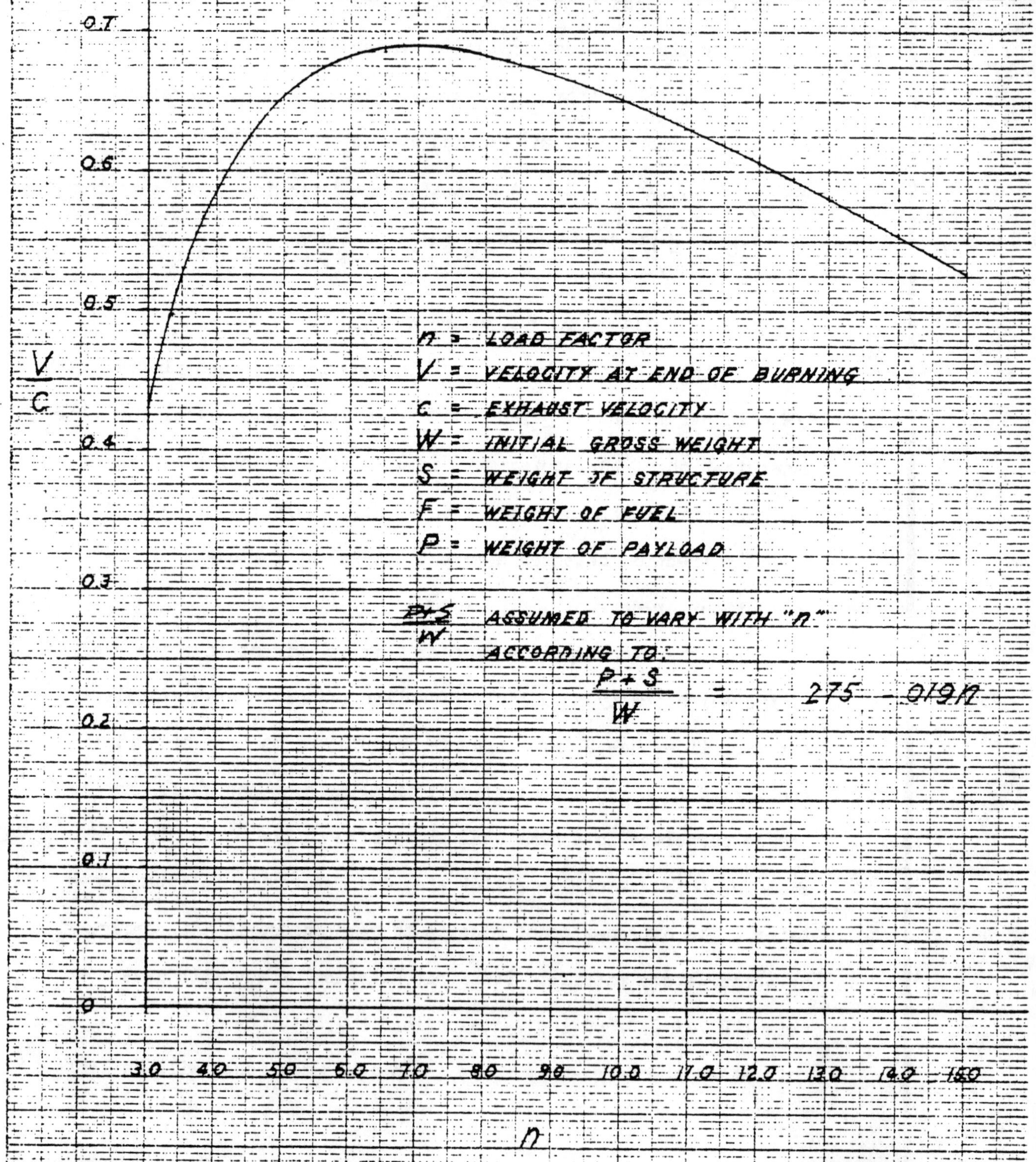

n = LOAD FACTOR
V = VELOCITY AT END OF BURNING
C = EXHAUST VELOCITY
W = INITIAL GROSS WEIGHT
S = WEIGHT OF STRUCTURE
F = WEIGHT OF FUEL
P = WEIGHT OF PAYLOAD

$\frac{P+S}{W}$ ASSUMED TO VARY WITH "n"
ACCORDING TO.

$$\frac{P+S}{W} = .275 - .019 n$$

FIGURE I
CHAPTER 2

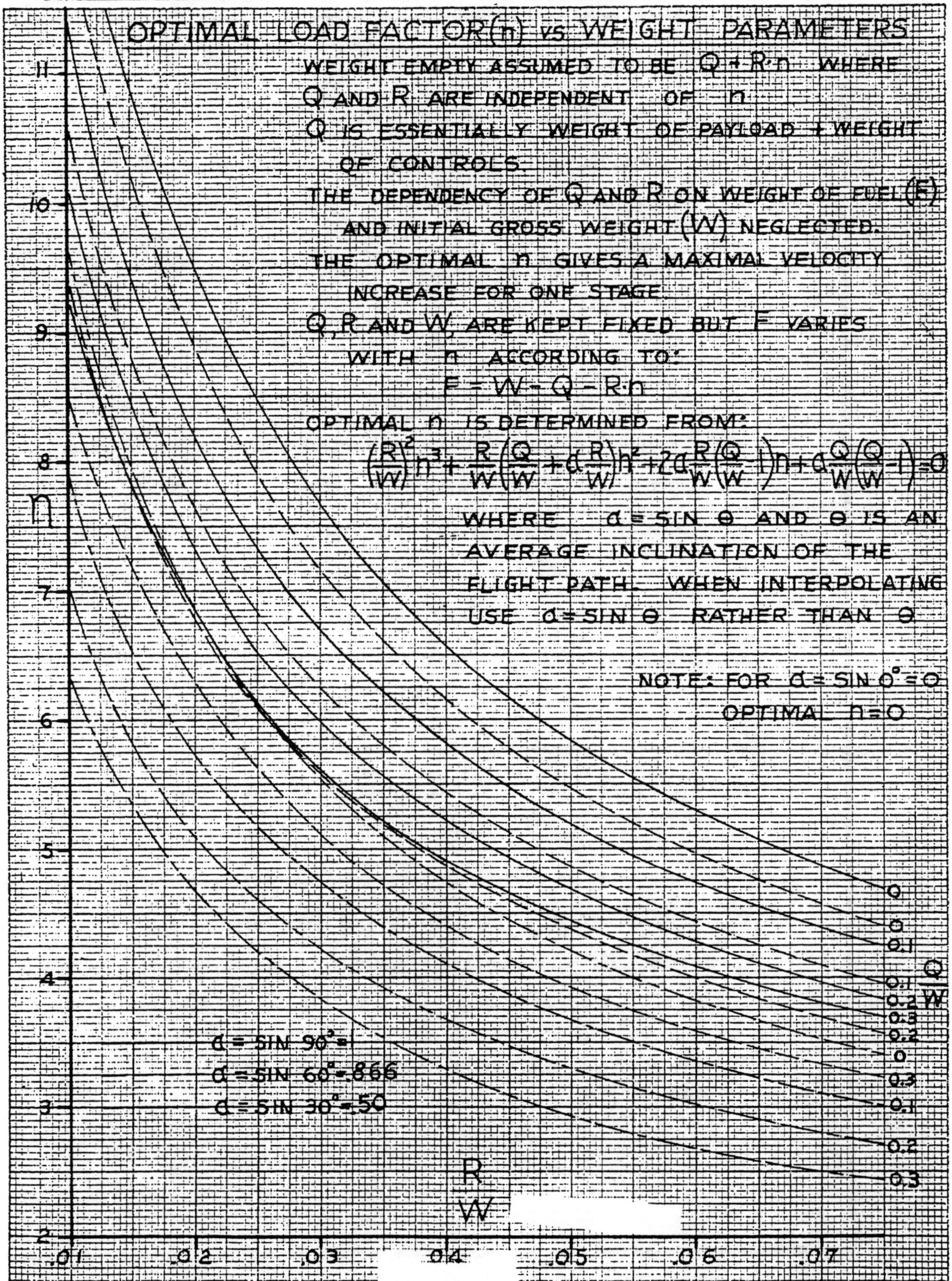

FORM 23.95
(REV. 7-42)

Analysis PRELIMINARY DESIGN OF SATELL

Prepared by E.S. Rutowski

Date 5-2-46

DOUGLAS AIRCRAFT COMPANY, INC.

Santa Monica   Plant

Page 85

Model 1033

Report No. SM11822

OPTIMAL LOAD FACTOR (n) vs WEIGHT PARAMETERS

WEIGHT EMPTY ASSUMED TO BE $Q + R \cdot n$ WHERE

Q AND R ARE INDEPENDENT OF n

Q IS ESSENTIALLY WEIGHT OF PAYLOAD + WEIGHT OF CONTROLS

THE DEPENDENCY OF Q AND R ON WEIGHT OF FUEL (F) AND INITIAL GROSS WEIGHT (W) NEGLECTED.

THE OPTIMAL n GIVES A MAXIMAL VELOCITY INCREASE FOR ONE STAGE

Q, R AND W, ARE KEPT FIXED BUT F VARIES WITH n ACCORDING TO:

$$F = W - Q - R \cdot n$$

OPTIMAL n IS DETERMINED FROM:

$$\left(\frac{R}{W}\right)^2 n^3 + \frac{R}{W}\left(\frac{Q}{W} + a\frac{R}{W}\right)n^2 + 2a\frac{R}{W}\left(\frac{Q}{W} - 1\right)n + a\frac{Q}{W}\left(\frac{Q}{W} - 1\right) = 0$$

WHERE $a = \sin \theta$ AND θ IS AN AVERAGE INCLINATION OF THE FLIGHT PATH. WHEN INTERPOLATING USE $a = \sin \theta$ RATHER THAN θ

NOTE: FOR $a = \sin 0° = 0$
OPTIMAL $n = 0$

$a = \sin 90° = 1$
$a = \sin 60° = .866$
$a = \sin 30° = .50$

$\frac{Q}{W}$

0
0
0.1
0.1
0.2
0.3
0.2
0
0.3
0.1
0.2
0.3

$\frac{R}{W}$

.01   .02   .03   .04   .05   .06   .07

FIGURE 2 CHAPTER 8

FORM 28-2-1
(REV. 8-42)

PREPARED BY: F. A. Lagerstrom DOUGLAS AIRCRAFT COMPANY, INC.   PAGE: 80

DATE: May 2, 1946 (Corr. 5-27-6) SANTA MONICA _____ PLANT   MODEL #1033

TITLE: PRELIMINARY DESIGN OF SATELLITE VEHICLE   REPORT NO. SM-11827

Chapter 8

constant, but the load factor will be varied from its design value 6.5.

The analysis given in Chapter 7 shows that the weight of the structure S_1 will vary with the load factor n according to the equation

$$(11) \quad S_1 = 14,300 \left(\frac{F_1}{140,000}\right)^{4/3} \frac{n}{6.5} + 7,780 \times \frac{55.2}{t_B} \left(\frac{F_1}{140,000}\right)$$

$$+ 11,400 \left(\frac{W_1}{233,669}\right)^{7/6} + \frac{3.0}{100} (233,669) \frac{n}{6.5}$$

The terms on the right hand side represent in order the weight of tanks, power plant, controls and various miscellaneous weights. The design values of these weights are 14,800 lbs., 7,780 lbs, 11,400 lbs and 3.0% of W_1 respectively. 55.2 and 6.5 are the design values of t_B and n. If we select a value of I of 240[*] sec and put W_1 and W_2 equal to their design values, then every term on the right hand side is a function of n only. The other variables may be eliminated with the aid of the equations $W_1 = S_1 + F_1 + W_2$ and $t_B = \frac{I \times F_1}{n(W_1 - F_1)}$. The result of this rather cumbersome numerical calculation is given in Figure 3. Once F_1, S_1 and t_B are known as functions of n, the final velocity V_F may be computed from (2) (with c = 32.2 x 240 = 7,750 fps). The result has been plotted as curve 1 in Figure 4 (the other graphs in Figure 4 will be discussed shortly). The optimal value of n is seen to be 7.5; V_F does not fall more than one percent below its maximum if n is kept between 6.5 and 8.75.

[*]This represents an average value during first stage.

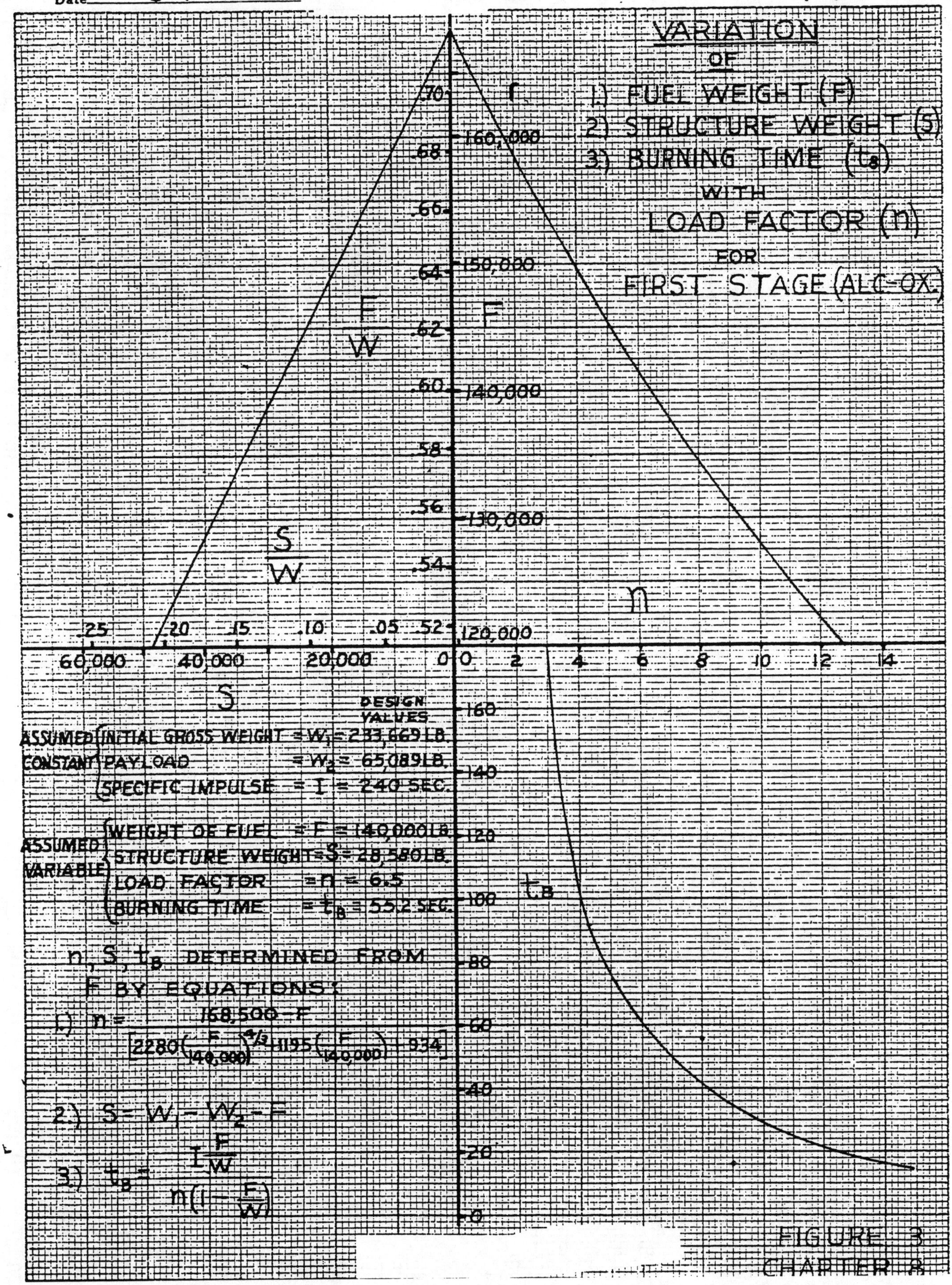

FORM 25 BS
(REV. 7 42)

Analysis PRELIMINARY DESIGN OF SATELLITE VEHICLE

Prepared by E.S. Rutowski

Date 5-1-46

DOUGLAS AIRCRAFT COMPANY, INC.

Santa Monica   Plant

Page 87

Model 1033

Report No. 11827

VARIATION
OF
1) FUEL WEIGHT (F)
2) STRUCTURE WEIGHT (S)
3) BURNING TIME (t_B)
WITH
LOAD FACTOR (n)
FOR
FIRST STAGE (ALC-OX.)

$\frac{F}{W}$ F

$\frac{S}{W}$

n

.70
.68 — 160,000
.66
.64 — 150,000
.62
.60 — 140,000
.58
.56 — 130,000
.54
.52 — 120,000

.25   .20   .15   .10   .05

60,000   40,000   20,000   0   0   2   4   6   8   10   12   14

S

DESIGN VALUES

160

ASSUMED (INITIAL GROSS WEIGHT = W_1 = 233,669 LB.

CONSTANT PAYLOAD   = W_2 = 65,089 LB.

(SPECIFIC IMPULSE = I = 240 SEC.

140

ASSUMED (WEIGHT OF FUEL = F = 140,000 LB.   120

VARIABLE (STRUCTURE WEIGHT = S = 28,580 LB.

(LOAD FACTOR = n = 6.5

(BURNING TIME = t_B = 55.2 SEC.   100 t_B

n, S, t_B DETERMINED FROM   80

F BY EQUATIONS:

1) $n = \dfrac{168,500 - F}{\left[2280\left(\dfrac{F}{140,000}\right)^{4/3} + 1195\left(\dfrac{F}{140,000}\right) - 934\right]}$ 60

2) $S = W_1 - W_2 - F$ 40

3) $t_B = \dfrac{I \dfrac{F}{W}}{n\left(1 - \dfrac{F}{W}\right)}$ 20

0

FIGURE 3

CHAPTER II

FORM 25 BS
(REV. 7-42)

Analysis PRELIMINARY DESIGN OF SATELLITE VEHICLE
Prepared by E.S.Rutowski
Date 4-29-46

DOUGLAS AIRCRAFT COMPANY, INC.
Santa Monica

Page 88
Model 1033
Plant
Report No. 11827

TOTAL ENERGY & FINAL VELOCITY
VS
LOAD FACTOR
FIRST STAGE (ALC.-OX.)

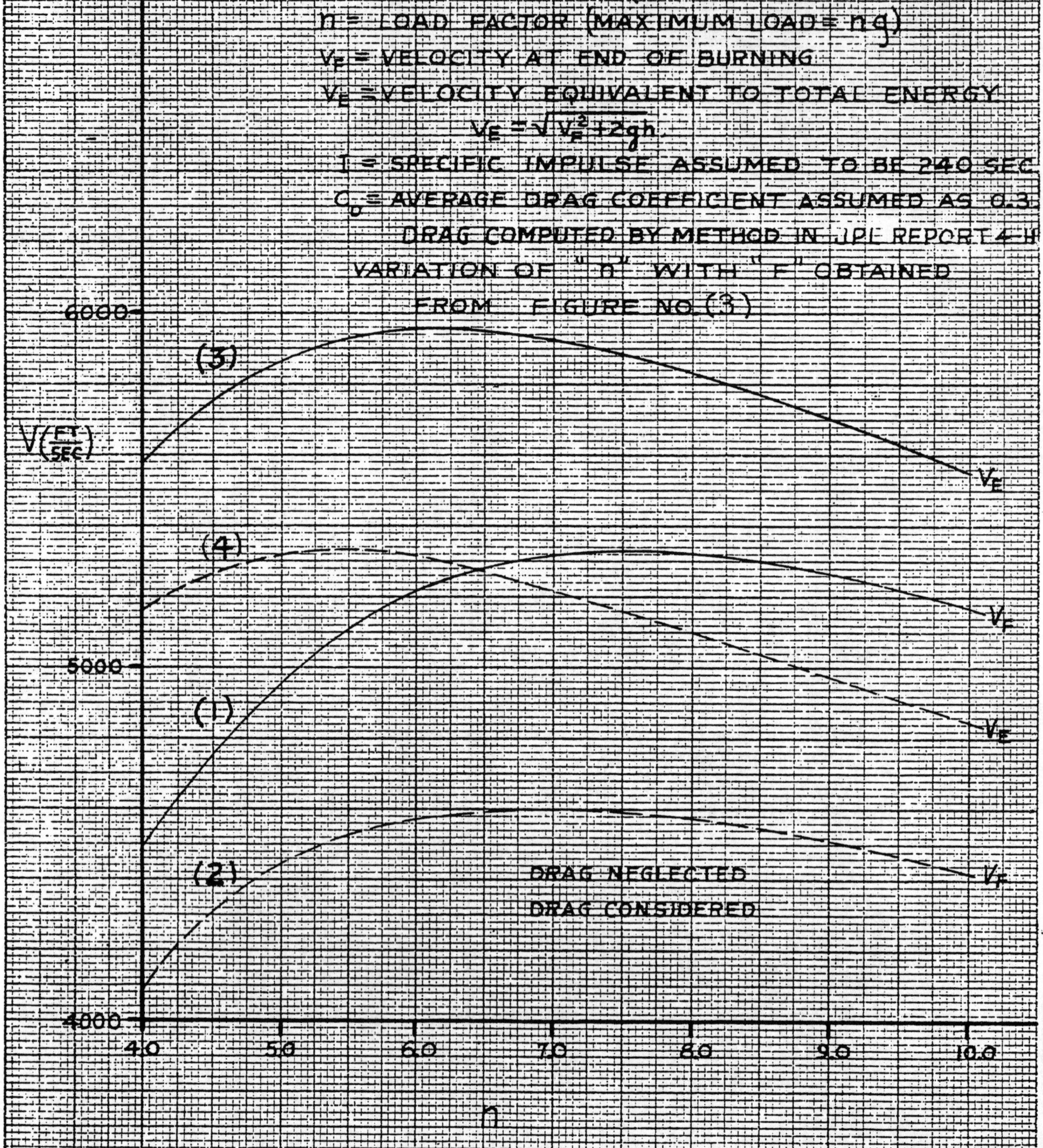

n = LOAD FACTOR (MAXIMUM LOAD = ng)
V_F = VELOCITY AT END OF BURNING
V_E = VELOCITY EQUIVALENT TO TOTAL ENERGY

$$V_E = \sqrt{V_F^2 + 2gh}$$

I = SPECIFIC IMPULSE ASSUMED TO BE 240 SEC
C_D = AVERAGE DRAG COEFFICIENT ASSUMED AS 0.3
DRAG COMPUTED BY METHOD IN JPL REPORT 4-H
VARIATION OF "n" WITH "F" OBTAINED
FROM FIGURE NO. (3)

$V\left(\frac{FT}{SEC}\right)$

6000 —

5000 —

4000 —

DRAG NEGLECTED
DRAG CONSIDERED

(1) (2) (3) (4)

V_E V_F V_E V_F

4.0   5.0   6.0   7.0   8.0   9.0   10.0

n

FIGURE 4
CHAPTER 8

Chapter 8

Drag - For a rapid estimate of the effect of drag on performance, the method of successive approximations is recommended. The equation for a vertical trajectory is (D denotes drag):

$$(12) \quad \frac{dV}{dt} = -\frac{c}{m}\frac{dm}{dt} - g - \frac{D}{m}.$$

A zero order solution is obtained by putting $D = 0$. Then (12) reduces to (1) and by integration one obtains the vacuum trajectory, i.e., velocity and altitude as functions of time. Using these functions, $\frac{D}{m}$ may be expressed as a function of time. The right hand side then depends on time only, and by integrating one obtains the first order solution for V as a function of time. The method has been elaborated in JPL-GALCIT Report No. 4-11, "Vertical Flight Performance of Rocket Missiles" by W. Z. Chien. It has been applied to the example discussed above (first stage) and the result is plotted as curve 2 in Figure 4. It can be seen that drag moves the maximum from $n=7.5$ to $n=7.0$.

Total Energy - So far, we have measured performance in terms of the final velocity V_F. However, it would seem that one really should consider the total energy gained which consists of kinetic + potential energy. This total energy may conveniently be represented by an "equivalent velocity" V_E, defined by $V_E = \sqrt{V_F^2 + 2gh}$ where h is the altitude gained. The equivalent velocity has also been plotted with and without drag as curves 3 and 4 in Figure 4. The values of n for maximum V_F are 6.0 (vacuum) and 5.5, as compared with the values 7.5 and 7.0 for maximum V_F.

If our vehicle actually were a single stage rocket V_E would clearly be the significant value. However, if it is one of the initial stages of a multi-stage rocket, the situation is more complicated. The reason

Chapter 8

is that a rocket motor is characterized by its thrust T which is essentially independent of flying conditions. Its power on the other hand is $T \times V$ and hence directly proportional to the speed of the vehicle, and the total energy gained during one stage is equal to the work done $= \int F ds = \int_{0}^{t_B} F \times V dt$ (neglecting drag and gravity). This shows the importance of the initial velocity. But the initial velocity of the second stage is the final velocity of the first stage. Thus by trying to get as much total energy as possible during the first stage one might lose out on later stages.

Returning to our specific example we conclude that the optimal n lies somewhere between the value for $V_E(7.0)$ and that for $V_F(5.5)$. Considering the flatness of the graphs, the design value $n = 6.5$ represents a reasonable compromise.

<u>Throttling</u> - On the basis of the above discussion of the influence of maximum acceleration on weight, it seems logical to investigate whether weight can be saved by throttling during the later portions of the burning period where the accelerations have increased considerably. However, a moment's reflection shows that this will not reduce the critical design condition on some of the structural components. Motors and pumps will have to be designed for maximal thrust rather than maximal acceleration and hence nothing is saved in the weight of the power plant. The load on the tanks of the first stage is greatest during the first part of the burning period when the tanks are full. Consequently, the weight of these tanks will be unaffected by throttling.

Chapter 8

Calculations have shown that if the maximum acceleration is reduced from 6.5g to 4g by throttling, the gain in structural weight is sufficiently greater than the loss in efficiency of the rocket, that a gain of about 3% on velocity is obtained. However, it is felt that this gain is not sufficient to warrant the additional complication in the present study.

Chapter 8

Proportioning of Stages

Multi-stage Vehicle - In Chapter 5 the concept of a multi-stage rocket was introduced. A simplified analysis showed that if the structural weight ratio is the same for all stages the performance is a maximum if the payload and the weights of the various stages form a geometric progression:

$$(11) \quad \frac{W_n}{P} = \frac{W_{n-1}}{W_n} = \cdots \frac{W_1}{W_2}$$

The final velocity at the end of burning of the nth stage is thus

$$(12) \quad \frac{V_F}{c} = -n\ln\left(\frac{S}{W} + \left(\frac{P}{W}\right)\frac{1}{n}\right)$$

This basic formula was plotted in Chapter 5 for $n = 1,2,3,4,5$ and $\frac{S}{W} = .16$. On the next pages four additional charts of this type are given for $\frac{S}{W} = .1, .143, .182$ and $.25$ (Figures 5A - 5D). The value .182 was achieved by the Germans in their later redesign of the V-2.

We now turn to a detailed consideration of the problem of optimal proportioning of the stages, when more realistic assumptions are introduced. It will turn out that factors like inclination, variable exhaust velocity, etc., will cause deviations of the optimal proportions given by the geometric series. However, these deviations will in general be small, sometimes insignificant. It will also be seen that the maxima

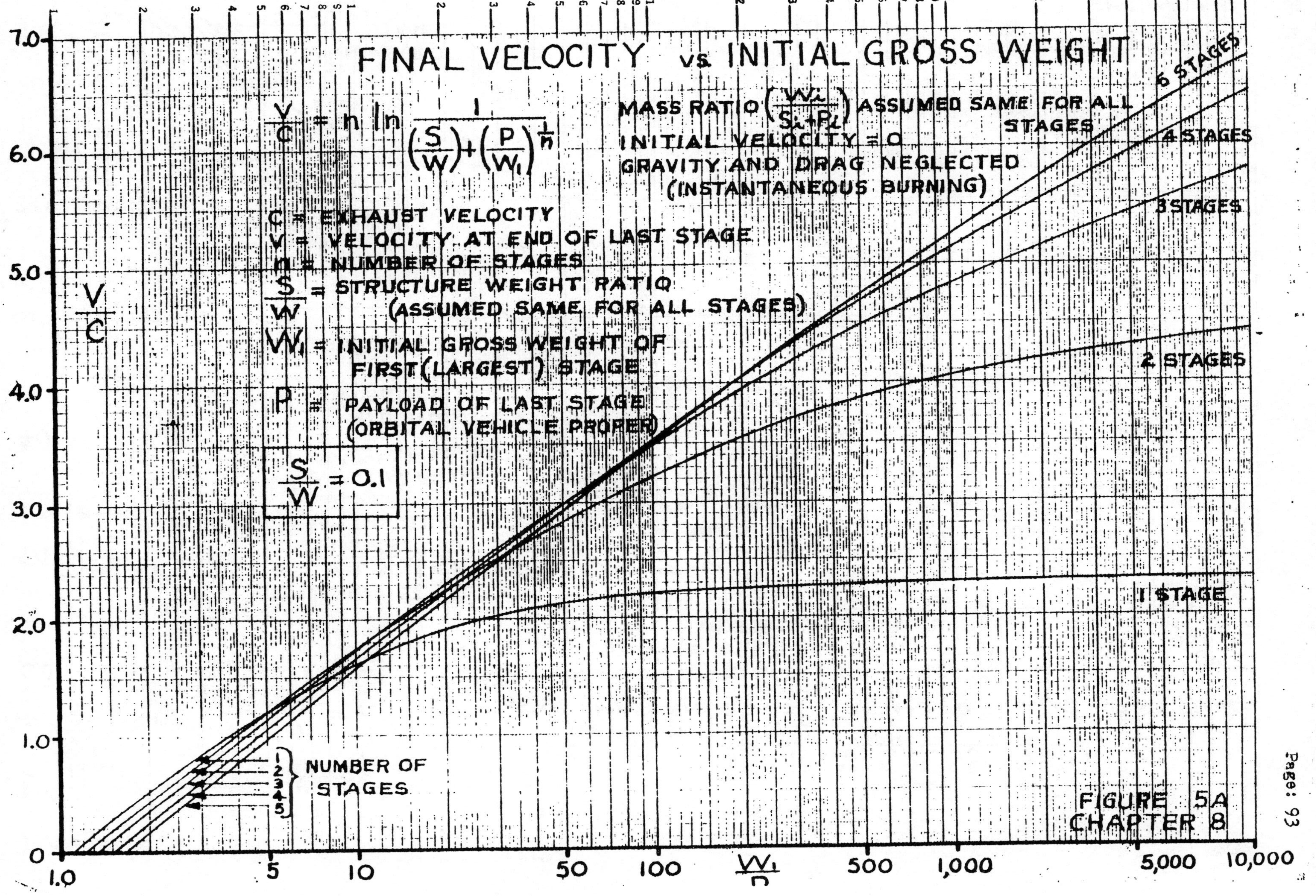
FINAL VELOCITY vs INITIAL GROSS WEIGHT
V/C = n ln 1/((S/W)+(P/W1)^(1/n))
MASS RATIO (Wi/(Si+PL)) ASSUMED SAME FOR ALL STAGES
INITIAL VELOCITY = 0
GRAVITY AND DRAG NEGLECTED
(INSTANTANEOUS BURNING)
C = EXHAUST VELOCITY
V = VELOCITY AT END OF LAST STAGE
n = NUMBER OF STAGES
S/W = STRUCTURE WEIGHT RATIO
(ASSUMED SAME FOR ALL STAGES)
W1 = INITIAL GROSS WEIGHT OF FIRST (LARGEST) STAGE
P = PAYLOAD OF LAST STAGE
(ORBITAL VEHICLE PROPER)
S/W = 0.1
6 STAGES
4 STAGES
3 STAGES
2 STAGES
1 STAGE
1 2 3 4 5 NUMBER OF STAGES
V/C
W1/n
7.0
6.0
5.0
4.0
3.0
2.0
1.0
0
1.0
5
10
50
100
500
1,000
5,000
10,000
FIGURE 5A
CHAPTER 8

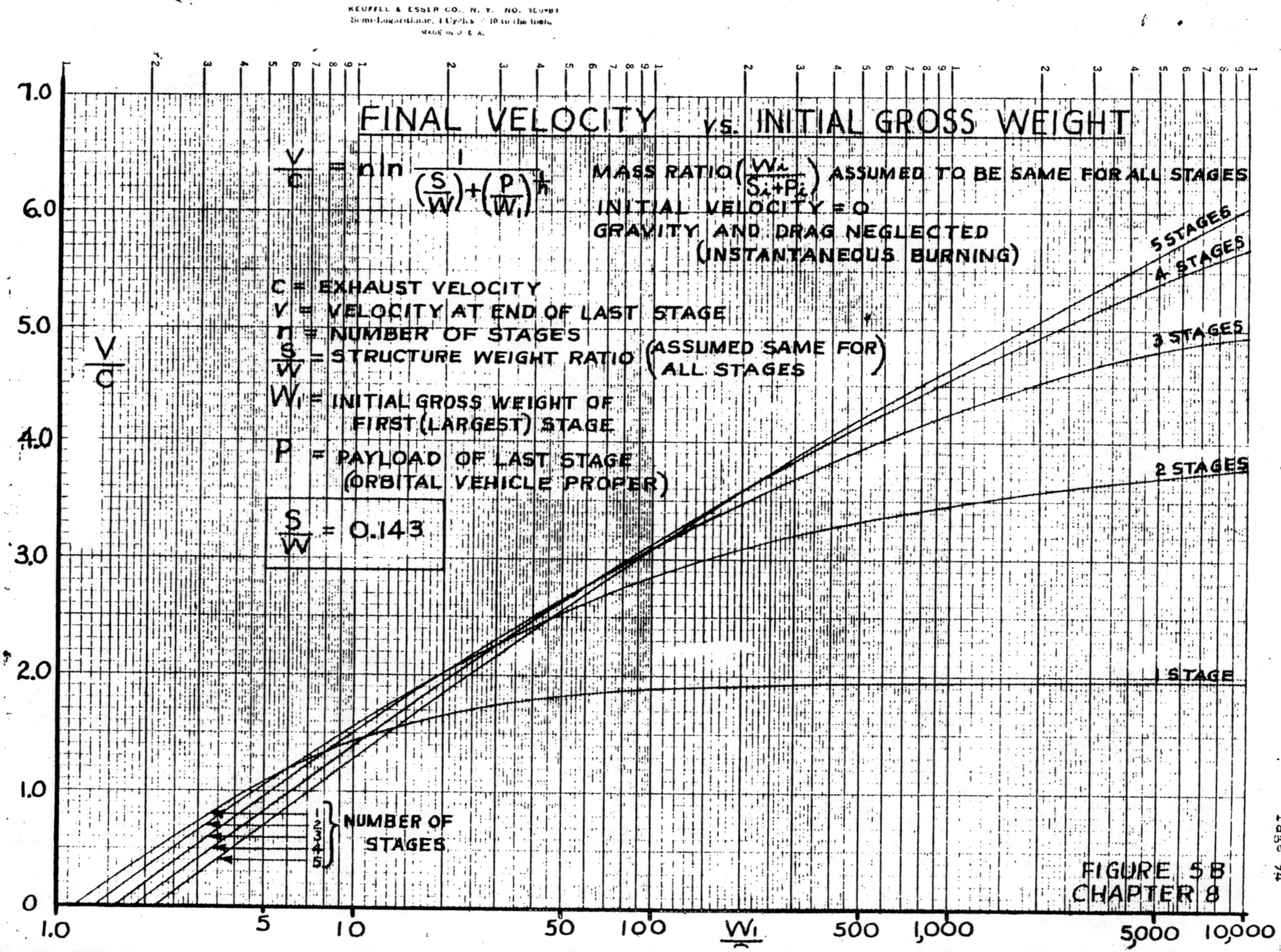
KEUFFEL & ESSER CO., N. Y. NO. 359-81
Semi-Logarithmic, 4 Cycles × 10 to the inch
MADE IN U. S. A.
FINAL VELOCITY vs. INITIAL GROSS WEIGHT
V/C = n ln 1/((S/W)+(P/W1)^(1/n))
MASS RATIO (Wi/(Si+Pi)) ASSUMED TO BE SAME FOR ALL STAGES
INITIAL VELOCITY = 0
GRAVITY AND DRAG NEGLECTED
(INSTANTANEOUS BURNING)
C = EXHAUST VELOCITY
V = VELOCITY AT END OF LAST STAGE
n = NUMBER OF STAGES
S/W = STRUCTURE WEIGHT RATIO (ASSUMED SAME FOR ALL STAGES)
W1 = INITIAL GROSS WEIGHT OF FIRST (LARGEST) STAGE
P = PAYLOAD OF LAST STAGE (ORBITAL VEHICLE PROPER)
S/W = 0.143
5 STAGES
4 STAGES
3 STAGES
2 STAGES
1 STAGE
NUMBER OF STAGES
1
2
3
4
5
V/C
7.0
6.0
5.0
4.0
3.0
2.0
1.0
0
W1/P
1.0
5
10
50
100
500
1,000
5,000
10,000

FIGURE 5B
CHAPTER 8

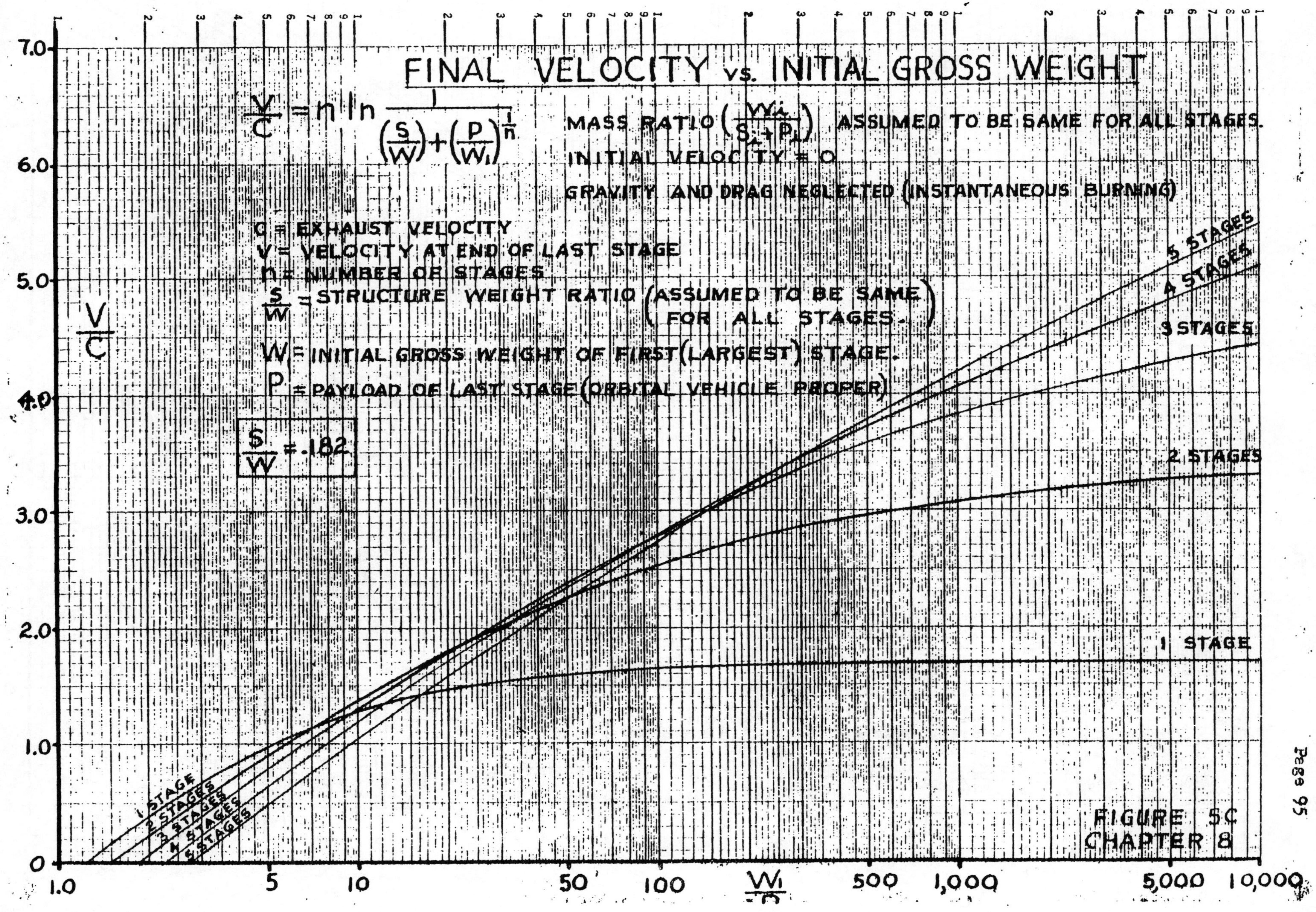
KEUFFEL & ESSER CO., N. Y. NO. 359-81
Semi-Logarithmic, 4 Cycles × 10 to the inch
MADE IN U.S.A.
FINAL VELOCITY vs. INITIAL GROSS WEIGHT
$\frac{V}{C} = n \ln \frac{1}{\left(\frac{S}{W}\right) + \left(\frac{P}{W_1}\right)^{\frac{1}{n}}}$
MASS RATIO $\left(\frac{W_i}{S_i + P_i}\right)$ ASSUMED TO BE SAME FOR ALL STAGES.
INITIAL VELOCITY = 0
GRAVITY AND DRAG NEGLECTED (INSTANTANEOUS BURNING)
C = EXHAUST VELOCITY
V = VELOCITY AT END OF LAST STAGE
n = NUMBER OF STAGES
$\frac{S}{W}$ = STRUCTURE WEIGHT RATIO (ASSUMED TO BE SAME FOR ALL STAGES.)
W = INITIAL GROSS WEIGHT OF FIRST (LARGEST) STAGE.
P = PAYLOAD OF LAST STAGE (ORBITAL VEHICLE PROPER)
$\frac{S}{W} = .182$
$\frac{V}{C}$
$\frac{W_1}{P}$
1 STAGE
2 STAGES
3 STAGES
4 STAGES
5 STAGES
FIGURE 5C
CHAPTER 8

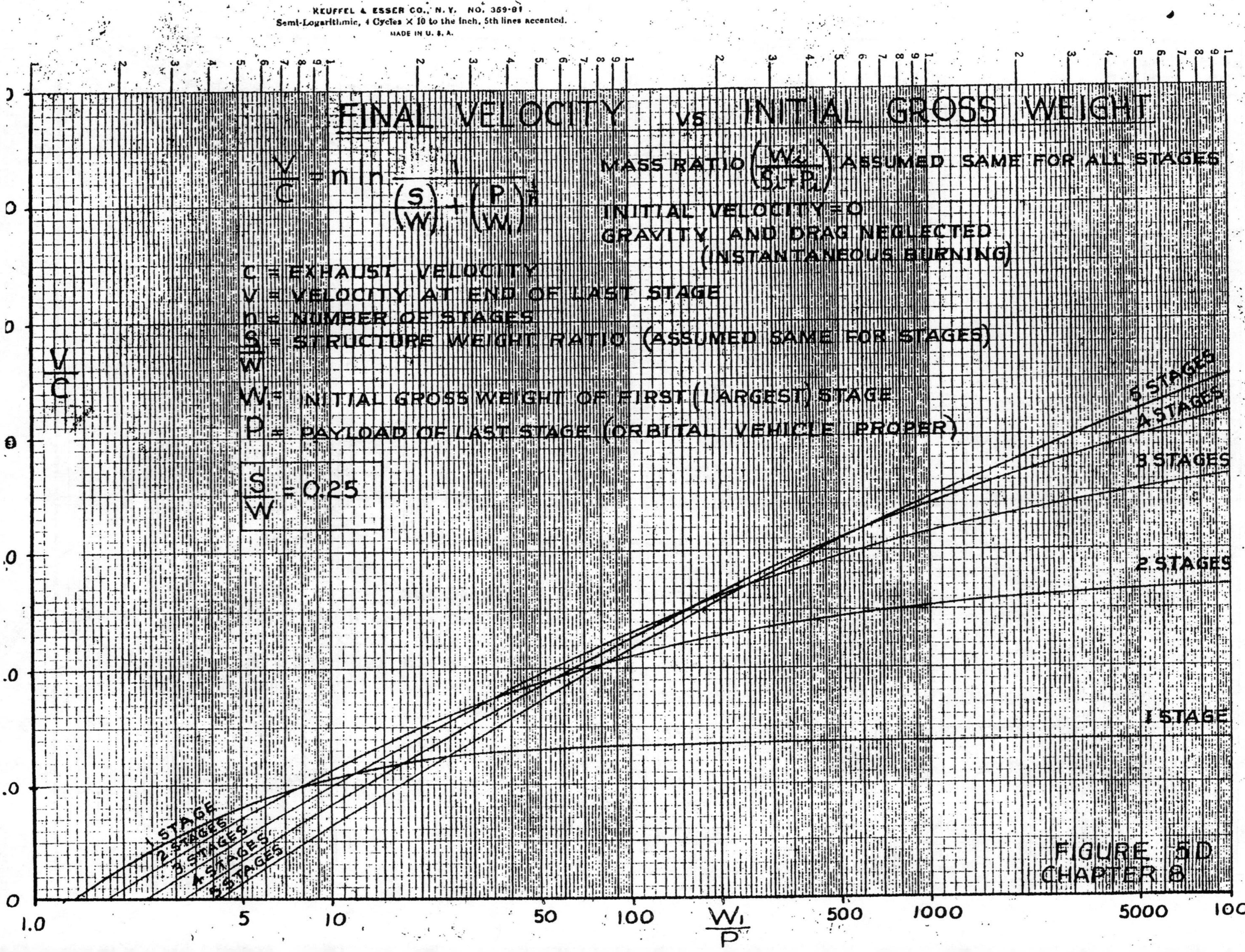
KEUFFEL & ESSER CO., N.Y. NO. 359-81
Semi-Logarithmic, 4 Cycles × 10 to the Inch, 5th lines accented.
MADE IN U.S.A.
FINAL VELOCITY VS INITIAL GROSS WEIGHT
$\frac{V}{C} = n \ln \frac{1}{\left(\frac{S}{W}\right) + \left(\frac{P}{W_1}\right)^{\frac{1}{n}}}$
MASS RATIO $\left(\frac{W_i}{S_i + P_i}\right)$ ASSUMED SAME FOR ALL STAGES
INITIAL VELOCITY = 0
GRAVITY AND DRAG NEGLECTED
(INSTANTANEOUS BURNING)
C = EXHAUST VELOCITY
V = VELOCITY AT END OF LAST STAGE
n = NUMBER OF STAGES
$\frac{S}{W}$ = STRUCTURE WEIGHT RATIO (ASSUMED SAME FOR STAGES)
W_1 = INITIAL GROSS WEIGHT OF FIRST (LARGEST) STAGE
P = PAYLOAD OF LAST STAGE (ORBITAL VEHICLE PROPER)
$\frac{S}{W} = 0.25$
$\frac{V}{C}$
$\frac{W_1}{P}$
1.0
5
10
50
100
500
1000
5000
10000
1 STAGE
2 STAGES
3 STAGES
4 STAGES
5 STAGES
FIGURE 5D
CHAPTER 8

Chapter 8

in general are fairly flat i.e. deviations from the theoretical maxima will cause but small loss in performance. In order to bring out the trends clearly, the various factors will be considered separately. When concrete numerical computations are made a two-stage rocket will be considered. If values are desired for rockets having a great number of stages, these results may be applied to the stages in pairs. The following formulation of the problem applies to all such two-stage rockets: The gross weight of the first stage (W_1) and the pay load of the second stage (P) will be fixed. The weight of the second stage (W_2) will be varied and the performance plotted against W_2 (or some equivalent variable like $\frac{W_2}{P}$). Quantities like n, c, $\frac{S}{W}$ will also be considered fixed except when otherwise stated. They may or may not be the same for both stages, depending on what factor is being studied.

Gravity. Inclination - If gravity is considered the optimal proportioning of the stages is still a geometrical progression as long as performance is being gauged by the final velocity V_F. (We thereby assume that the maximal load factor n rather than the burning time t_B is kept the same for both stages.) This result may be proved by standard methods of calculus or by a qualitative method to be discussed later in this chapter. A plot for an actual concrete case (Figure 6, curve 1) shows this maximum to be rather flat. If the equivalent velocity V_E is considered instead, a small displacement of the maximum to the right takes place as shown by curve 2.

In the same figure the effect of inclination is shown by curves 3, 4 and 5. The first stage is assumed vertical. If the second stage has an

EFFECT OF GRAVITY AND INCLINATION
ON
OPTIMAL PROPORTIONING OF STAGES OF TWO-STAGE ROCKET

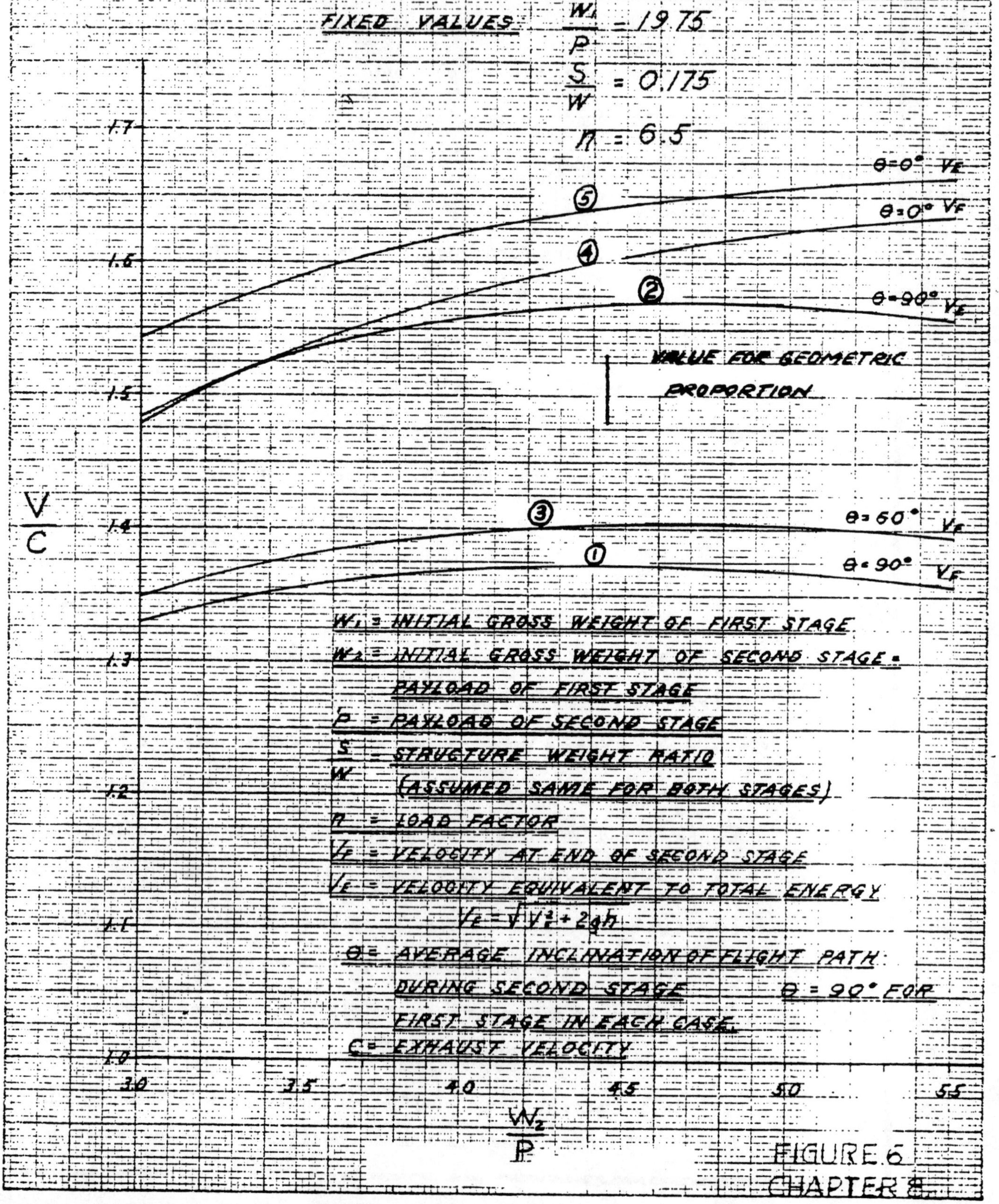

FIXED VALUES

$$\frac{W_1}{P} = 19.75$$

$$\frac{S}{W} = 0.175$$

$$n = 6.5$$

$\theta = 0°$ V_E

$\theta = 0°$ V_F

$\theta = 90°$ V_E

VALUE FOR GEOMETRIC PROPORTION

$\theta = 60°$ V_F

$\theta = 90°$ V_F

$\frac{V}{C}$ (vertical axis: 1.0 to 1.7)

Horizontal axis $\frac{W_2}{P}$: 3.0, 3.5, 4.0, 4.5, 5.0, 5.5

W_1 = INITIAL GROSS WEIGHT OF FIRST STAGE

W_2 = INITIAL GROSS WEIGHT OF SECOND STAGE = PAYLOAD OF FIRST STAGE

P = PAYLOAD OF SECOND STAGE

$\frac{S}{W}$ STRUCTURE WEIGHT RATIO (ASSUMED SAME FOR BOTH STAGES)

n = LOAD FACTOR

V_E = VELOCITY AT END OF SECOND STAGE

V_F = VELOCITY EQUIVALENT TO TOTAL ENERGY

$$V_F = \sqrt{V_E^2 + 2gh}$$

θ = AVERAGE INCLINATION OF FLIGHT PATH DURING SECOND STAGE $\theta = 90°$ FOR FIRST STAGE IN EACH CASE.

C = EXHAUST VELOCITY

FIGURE 6

CHAPTER 8

average inclination with the horizontal of $60°$, a rather insignificant displacement of the maximum to the right takes place (curve 3). In order to bring out the trend more clearly, the exaggerated case of a horizontal second stage was considered (curves 4 and 5). In this case, the displacement to the right of the maximum of both V_F and V_E is considerable.

Exhaust Velocity - Due to altitude effect on power plants the average exhaust velocity of the second stage is usually larger than that of the first stage (a 15% increase is a typical value). Multi-stage rockets have also been proposed where the second stage employe fuel with considerably higher exhaust velocity than that of the first stage.

For a two-stage rocket, the following formula for the optimal proportioning was obtained by calculus:

$$(13) \quad \frac{W_2}{\sqrt{PW_1}} = \frac{1}{2r} \left(\frac{W}{S}\right)\left[-\sqrt{\frac{P}{W_1}}(r-1) + \sqrt{\frac{P}{W_1}(r-1)^2 + 4\left(\frac{S}{W}\right)^2 r}\right] \quad \text{where } r = \frac{C_2}{C_1}$$

For values of r near one, this formula can be linearized to give

$$(14) \quad \frac{W_2}{\sqrt{PW_1}} = 1 - \frac{1}{2}(r-1)\left[1 + \frac{W}{S}\sqrt{\frac{P}{W_1}}\right]$$

Plots of values obtained from these equations are given in Figures 7A and 7B. For the typical values of $\left(\frac{S}{W}\right) = .143$ and $\frac{W_1}{P} = 20$ and the exhaust velocity of the second stage 15% larger than that of the first stage, the optimal value of W_2 is about 16% smaller than the geometric mean of P and W_1.

Analysis_____

Prepared b, L.S. Rudonski   DOI   VY. INC.

Date   5-3-46   Santa Monica   Plant

Page 100
Model 1033
Report No. 11827

OPTIMAL PROPORTIONING OF STAGES OF TWO-STAGE ROCKET
WHEN
EXHAUST VELOCITIES OF THE TWO STAGES ARE DIFFERENT

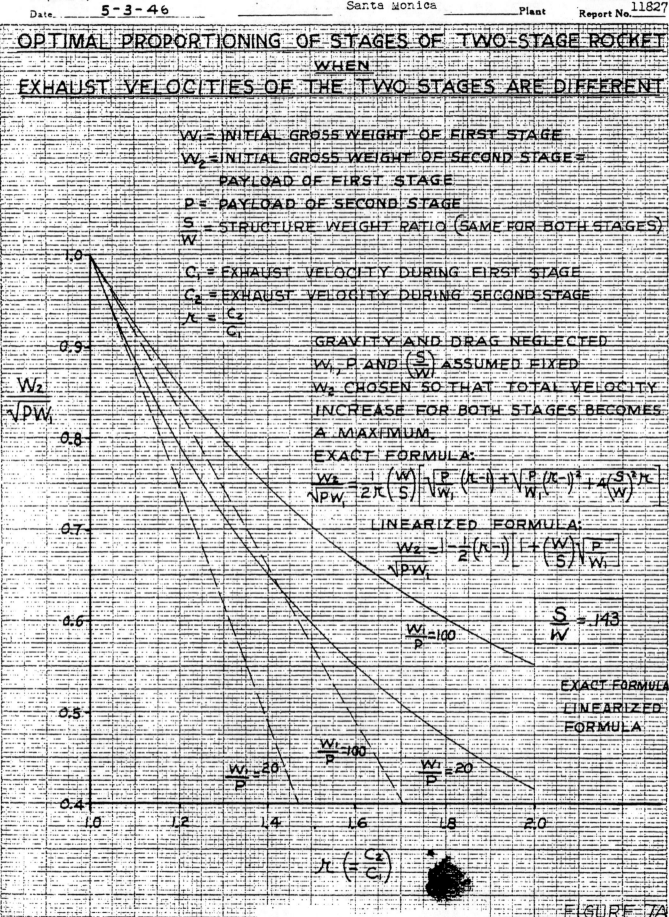

W_1 = INITIAL GROSS WEIGHT OF FIRST STAGE

W_2 = INITIAL GROSS WEIGHT OF SECOND STAGE =

PAYLOAD OF FIRST STAGE

P = PAYLOAD OF SECOND STAGE

$\dfrac{S}{W}$ = STRUCTURE WEIGHT RATIO (SAME FOR BOTH STAGES)

C_1 = EXHAUST VELOCITY DURING FIRST STAGE

C_2 = EXHAUST VELOCITY DURING SECOND STAGE

$\pi = \dfrac{C_2}{C_1}$

GRAVITY AND DRAG NEGLECTED

W_1, P AND $\left(\dfrac{S}{W}\right)$ ASSUMED FIXED

W_2 CHOSEN SO THAT TOTAL VELOCITY

INCREASE FOR BOTH STAGES BECOMES

A MAXIMUM.

EXACT FORMULA:

$$\frac{W_2}{\sqrt{PW_1}} = \frac{1}{2\pi}\left(\frac{W}{S}\right)\left[\sqrt{\frac{P}{W_1}}(\pi-1) + \sqrt{\frac{P}{W_1}(\pi-1)^2 + 4\left(\frac{S}{W}\right)^2 \pi}\right]$$

LINEARIZED FORMULA:

$$\frac{W_2}{\sqrt{PW_1}} = 1 - \frac{1}{2}(\pi-1)\left[1 + \left(\frac{W}{S}\right)\sqrt{\frac{P}{W_1}}\right]$$

$\dfrac{S}{W} = .143$

$\dfrac{W_1}{P} = 100$

EXACT FORMULA

LINEARIZED
FORMULA

$\dfrac{W_1}{P} = 20$

$\dfrac{W_1}{P} = 100$

$\dfrac{W_1}{P} = 20$

$\pi \left(= \dfrac{C_2}{C_1}\right)$

$\dfrac{W_2}{\sqrt{PW_1}}$

FIGURE 7A
CHAPTER 3

Analysis PRELIMINARY DESIGN OF S.   E   CRAFT COMPANY, INC.   Page 101
Prepared E.S.Rutowski   D   Model 1033
Date 5-3-46   Santa Monica   Plant   Report No. 11827

OPTIMAL PROPORTIONING OF STAGES OF TWO-STAGE ROCKET

WHEN

EXHAUST VELOCITIES OF THE TWO STAGES ARE DIFFERENT

(USING LINEARIZED FORMULA)

W_1 = INITIAL GROSS WEIGHT OF FIRST STAGE

W_2 = INITIAL GROSS WEIGHT OF SECOND STAGE = PAYLOAD OF FIRST STAGE

P = PAYLOAD OF SECOND STAGE

$\frac{S}{W}$ = STRUCTURE WEIGHT RATIO (SAME FOR BOTH STAGES)

C_1 = EXHAUST VELOCITY DURING FIRST STAGE

C_2 = EXHAUST VELOCITY DURING SECOND STAGE

$\kappa = \frac{C_2}{C_1}$

$\frac{W_2}{\sqrt{PW}}$

$\frac{W_1}{P} = 100$

$\frac{W_1}{P} = 36$

$\frac{W_1}{P} = 20$

$\frac{W}{S}$

W_1, P AND $\left(\frac{S}{W}\right)$ ASSUMED FIXED

W_2 CHOSEN SO THAT TOTAL VELOCITY INCREASE FOR BOTH STAGES BECOMES A MAXIMUM

GRAVITY AND DRAG NEGLECTED

EXACT FORMULA:

$$\frac{W_2}{\sqrt{PW}} = \frac{1}{2\kappa}\left(\frac{W}{S}\right)\left[-\sqrt{\frac{P}{W_1}}(\kappa-1)+\sqrt{\frac{P}{W_1}(\kappa-1)^2+4\left(\frac{S}{W}\right)\kappa}\right]$$

SEE FIGURE (7A)

LINEARIZED FORMULA:

$$\frac{W_2}{\sqrt{PW_1}} = 1 - \frac{1}{2}(\kappa-1)\left[1+\left(\frac{W}{S}\right)\sqrt{\frac{P}{W_1}}\right]$$

$\kappa\left(=\frac{C_2}{C_1}\right)$

1.00   1.05   1.10   1.15   1.20   1.25

Chapter 8

Drag - An investigation of the effect of drag on the optimum proportioning of two stages was made for the case in which the first stage had higher drag than the second (this is the normal case encountered). The results showed that the influence of drag on optimal proportioning is comparatively insignificant.

General Considerations about Optimal Proportioning. Influence of Structural Weight Ratio - The following qualitative method for determining the optimal proportioning might sometimes prove useful. Details of the method are omitted.

Consider a two-stage rocket as before. Express the performance for each stage as a function of the payload weight ratio. Denote these functions by $f_1(x)$ and $f_2(x)$. Put $\frac{P}{W_1} = a = $ constant and the payload weight ratio of the first stage $= \frac{W_2}{W_1} = x$; then the payload weight ratio of the second stage is $\frac{P}{W_2} = \frac{a}{x}$. The total velocity increase expressed as a function of the payload weight ratio of the first stage is then $V(x) = f_1(x) + f_2\left(\frac{a}{x}\right)$

Consider the difference $d = V\left(\frac{a}{x}\right) - V(x) = \left[\left(f_2\left(\frac{a}{x}\right) - f_1\left(\frac{a}{x}\right)\right) - \left(f_2(x) - f_1(x)\right)\right]$. If d is zero, the graph of V against ░░░ a certain type of symmetry around the value where $\frac{a}{x} = x$, i.e. $x = \sqrt{a}$. It is then geometrically evident that this point of symmetry represents a maximum. The difference d is zero when f_1 and f_2 are equal or differ by a constant. This is true for the idealized case when $f_1(x) = f_2(x) = -c \ln\left(x + \frac{S}{W}\right)$. It remains true when gravity is considered.

Chapter 8

Now take the case of different exhaust velocities. Then the difference d is $= (\bar{c}_2 - \bar{c}_1)\left[\ln(x + \frac{S}{W}) - \ln(\frac{a}{x} + \frac{S}{W})\right]$ This is negative for $x < \frac{a}{x}$. From the geometry of the graph it then follows that the maximum has been displaced from $x = \sqrt{a}$ towards smaller x.

So far we have only verified previous findings. To obtain a new result, consider the case of variable structure weight ratio, e.g. $\frac{S_1}{W_1} > \frac{S_2}{W_2}$.

Then d = $c \ln \dfrac{\frac{S_1}{W_1} + \frac{a}{x}}{\frac{S_2}{W_2} + \frac{a}{x}} - c \ln \dfrac{\frac{S_1}{W_1} + x}{\frac{S_2}{W_2} + x}$.

This is also negative for $x < \frac{a}{x}$ and again we conclude that the maximum has been displaced towards the left, i.e., the second stage is smaller than the geometric mean when its structural weight ratio is less than that of the first stage.

Influence of the Various Stages on Each Other - Above we have sometimes applied the results obtained for a single stage vehicle to each separate phase of a multi-stage vehicle. However, the discussion of whether V_E or V_F is the significant performance parameter for the first stage of a multi-stage vehicle (see p. 89) showed that a certain amount of care is necessary.

Another example of how the various stages influence each other is this: For each separate stage one can find the optimal load factor by the methods described at the befinning of this chapter. However, each stage cannot be designed for its own optimal load factor. The reason is simply that any stage is subjected to the maximum loads of any previous stage. And in general the optimal

Chapter 8

load factor is lower for later stages.

<u>Application of General Methods to Actual Design</u>.- So far we have developed a general analysis of performance parameters. We now proceed to apply our results to actual designs of two vehicles. Our starting point will of course be the specifications of the orbital vehicle. Two quantities are revelant for our considerations: (1) orbital velocity; (2) weight of pay load.

Strictly the altitude of the orbit should likewise be given. However, it was found in Chapter 3 that the altitude had little effect on the energies required for various orbits. This is fortunate, because it implies that the shape of the trajectory will exert only a secondary influence on our choice of design proportions. A value 24,500 ft/sec was selected as the orbital velocity to be used in our present consideration. The pay load, selected on the basis of a reasonable allowance for scientific instruments, was taken as 500 lbs. However, for the purposes of the analysis in this chapter, we also have to count the 200 lbs. of "brains" as pay load. The reason is that this is a fixed weight item which occurs only in the last stage and which is not included in a normal estimate of the structure-weight ratio $\frac{S}{W}$. Thus, for the remainder of this chapter, the weight of the pay load will be considered to be 700 lbs.

Next we need a value of the parameter c (exhaust velocity) which specifies the performance of the power plant, in other words we have to select the fuel. Two different vehicles will be considered, one powered

PREPARED BY: P. A. Lagerstrom DOUGLAS AIRCRAFT COMPANY, INC. PAGE: 105
DATE: May 2, 1946 SANTA MONICA PLANT MODEL #1033
TITLE: PRELIMINARY DESIGN OF SATELLITE VEHICLE REPORT NO. SM 11827

Chapter 8

by oxygen-alcohol, the other by oxygen-hydrogen.

Alcohol-Oxygen Vehicle - If liquid oxygen and alcohol are used as propellants, the value c = 8,500 fps. is a reasonable average value for our initial work. In selecting this value consideration was given to the fact that most of the operation of the rocket will be at altitude. Using this value of c, we find that $\frac{V}{c}$ is 2.88. In the preliminary work we shall use a value of $\frac{S}{W}$ = .16 which is an average value obtained from the results of Chapter 7. For our present purpose it would be extremely convenient to have a correction factor to take care of these items. A good value of such a factor can only be based on long experience in designing, building and testing orbital vehicles. Since this experience is lacking, we use the following estimate. The V-2 was designed for a load factor of 6.5. A previous estimate in this chapter showed that with this load factor the losses due to gravity in a vertical trajectory amount to 25%. Inclination of the last portion reduces this to about 20%. Furthermore, preliminary calculations showed that drag will reduce the final velocity by 10%. Thus we arrive at a correction factor of 1.30. This increases the value of $\frac{V}{c}$ required from 2.88 to 3.74. A look at Figure 2 of Chapter 5 tells us that a single-stage rocket and a two-stage rocket are impossible. A three-stage rocket would have to have a weight ratio $\frac{W_1}{P}$ of 450 and a four-stage and a five-stage rocket would have about equal weight ratios of 330. Thus a three-stage vehicle would necessitate a gross weight of

Chapter   8

450 x 700 = 315,000 lbs. and a four-stage vehicle 330 x 700 =
230,000 lbs. This weight saving seems to warrant the additional
complications of having four stages instead of three, whereas
nothing is gained by adding a fifth stage. The weight of the other
stages will form a geometrical progression. The ratio of the weights
between two consecutive stages is $(330)^{1/4}$ = 4.27. At this stage
in the analysis, the detailed weight study of Chapter 7 was
undertaken. Hand in hand with this structural weight study, an
analysis of the optimum design proportions was made, using the
methods previously explained in this chapter. This combined study
resulted in the set of values presented in Chapter 7.

The next step in our design study will be to determine more
rigorously the actual trajectory of the vehicle, taking into account
the variation of drag, exhaust velocity and inclination. This
more detailed computation will be done in the next chapter. The
results obtained there will give us an indication of how accurate
the preliminary analysis of this chapter has been.

It is clear that it is impossible to maintain a strict logical
order in determining the proportions of the vehicle. Actually
one has to repeat the process described here several times, just
as when solving a problem by successive approximations. As a
"first approximation" for W_1 we found the value 230,000 lbs. above.
In the course of the revisions indicated above, this value was
changed to 233,669. Hence, we may conclude that the factor 1.30

Chapter 8

used above for correcting $\frac{V}{c}$ was a reasonably good estimate. The values for $\frac{S}{W}$ finally established were .168, .18, .157 and .156 for stages 1, 2, 3 and 4 respectively, which are likewise in satisfactory agreement with the assumed value of .16. It was unnecessary to alter the value of 6.5 for the maximum ratio of thrust to weight.

Hydrogen-Oxygen - When the analysis was first made for the liquid hydrogen-liquid oxygen rocket, the structural weight ratio was estimated to be 0.20. In addition, it was erroneously assumed that higher accelerations would be used than were used for the alcohol oxygen rocket with a consequent reduction in the correction factor to 1.2. Under these conditions, it appeared that a three stage hydrogen-oxygen rocket would give slightly smaller overall gross weights than a two-stage rocket. However, the gain was so small that it was decided to avoid the complications of the three-stage rocket and proceed with the design of a two-stage rocket.

As the work on structural weight analysis progressed, it became apparent that the acceleration would have to be reduced to the value used for the alcohol-oxygen rocket. Making allowance for this decreased acceleration and also for the higher drag of the hydrogen rocket, a revised correction factor of 1.32 was obtained. In addition, it was found that the structural weight

Chapter 8

ratio would increase to about 0.25. When these later figures were used, they showed that it was advantageous to use three stages instead of two. However, the design study for this vehicle had proceeded so far that it was inadvisable to alter the number of stages.

If we use an exhaust velocity of 13,500 ft/sec and a correction factor of 1.32 we find $\frac{V}{c} = 2.4$. For $\frac{S}{W} = 0.25$ figure 5D shows that a value of $\frac{W_1}{P}$ of 400 is necessary for a two-stage rocket. This corresponds to a gross weight of 280,000 lbs.

The final design values resulting from the combined structural weight and performance study are tabulated in Chapter 7. The overall gross weight was 291,564 lbs. The values of $\frac{W}{S}$ were .238 and .245 for the 1st and 2nd stages respectively.

It is worthwhile at this point to say a few words about the possibilities of a three-stage hydrogen-oxygen rocket. As mentioned above, the study had proceeded too far for alteration when the design values had crystallized sufficiently to show the definite advantages of using three stages. Let us examine this case at greater length. Using $\frac{V}{c} = 2.4$ and $\frac{S}{W} = 0.25$, we find from figure 5D that $\frac{W_1}{P} = 120$ for three stages. This implies that the overall gross weight of this vehicle would be 84,000 lbs. which is considerably less than the weight of either the two-stage hydrogen-oxygen rocket or the four-stage alcohol-oxygen rocket. From this we may conclude that the three stage liquid hydrogen-liquid oxygen should be given serious consideration in any further studies of satellite vehicles.

FORM 2S-S-1 (REV. 9-43)

PREPARED BY: F. H. Clauser / R. W. Krueger

DOUGLAS AIRCRAFT COMPANY, INC.

DATE: May 2, 1946   SANTA MONICA   PLANT

TITLE: PRELIMINARY DESIGN OF SATELLITE VEHICLE

PAGE: 110

MODEL: #1033

REPORT NO. SM-11827

Chapter 9

9. FINAL ESTABLISHMENT OF SIZES AND TRAJECTORIES

In the preceding chapter, we have discussed the choice of burning time for a single stage rocket and the optimum proportioning of weights between two successive stages for multi-stage rockets. In order to proceed farther in our analysis, it is necessary to have an integrated picture of the variation of altitude, speed, inclination and mass at all points along the trajectory. In order to obtain these data, we shall use the design values obtained in the last chapter and carry out detailed calculations of the entire trajectory. The results of these calculations will show how much these design values are in error and it will give us considerably more reliable values which could be used in repeating the design studies of chapter 8. Ideally, this process of iteration should be continued until a satisfactory set of final design data are obtained. For the present study, however, our attention was confined to an investigation of the trajectories for the two proposed designs, without a repetition of the calculations of chapter 8.

The vehicles which were selected for study and which served as a basis for the calculations of this chapter are tabulated below.

Vehicle Powered by Alcohol-Oxygen Rockets

Stage	1	2	3	4
Gross Wt. (lbs.)	233,669	53,689	11,829	2,868
Weight less fuel (lbs.)	93,669	21,489	4,729	1,148
Payload (lbs.)	53,689	11,829	2,868	500
Max. Diameter (in.)	157	138	105	90

Chapter 9

Vehicle Powered by Hydrogen-Oxygen Rockets

Stage	1	2
Gross Wt.	291,564	15,364
Weight Empty	84,564	4,464
Payload (lbs.)	15,364	500
Max. Diameter (in.)	248	167

In all of the vehicles, the rate of fuel consumption was maintained constant at a value calculated to give a maximum ratio of thrust to weight of 6.5. The drags were calculated according to the methods given in Appendix B. The variations of exhaust velocity with altitude were taken from the graphs in Chapter 6.

In order to avoid a mass of distracting details, none of the lengthy and involved trajectory calculations will be presented in this chapter. We shall simply indicate the methods used and present the final results that were obtained within the time available, which was, to say the least, insufficient to answer all the questions that will inevitably arise.

The mathematical developments necessary for this work, as well as samples of the calculation methods used, are presented in appendices C and D at the end of this report.

In the early stages of the work, we set for ourselves the goal of establishing a 500 lb. payload on an orbit approximately 100 miles above the surface of the earth, hoping it could remain there for a period of 5 to 10 days before its energy was dissipated in the rarefied atmosphere. The 500 lb. figure was chosen as a reasonable estimate of the weight of

PREPARED BY: E. H. Clauser
R. H. Krueger   DOUGLAS AIRCRAFT COMPANY, INC.   PAGE: 112

FORM 25-S-1
(REV. 8-43)

DATE: May 2, 1946   SANTA MONICA   PLANT   MODEL: #1033

TITLE: PRELIMINARY DESIGN OF SATELLITE VEHICLE   REPORT NO. SM-11827

Chapter 9

scientific apparatus necessary to obtain results sufficiently far reaching to make the undertaking worthwhile.

The altitude figure was a compromise between the desire to increase the altitude in order to reduce the drag and give larger limits of error in establishing the orbit on the one hand and to keep the orbital energies low and to aid the short wave radio control and tracking problem on the other. Preliminary calculations of energy dissipation in the orbit, based on conventional evaluation of drag coefficients and an isothermal atmosphere indicated that the vehicle could remain aloft for at least 5 to 10 days at an altitude of 100 miles. However, closer examination revealed that the conventional drag predictions were inadequate and that the assumption of an isothermal atmosphere was in error. Revised methods of drag prediction were developed and better estimates of the structure of the atmosphere* were obtained from an extensive search of the literature. The results of this revised study showed that it would be advisable to use a altitude of 300 to 400 miles in order to obtain the durations desired. The calculations were revised correspondingly but unfortunately time was not available for the work to be completed in detail.

In approaching the problem of launching a vehicle into a satellite orbit we see that consideration must be given to the following:

1. Obtaining time-velocity-acceleration-distance-mass-relationship for the trajectory used in placing the vehicle on a satellite orbit.

2. Devising a means of stably controlling the vehicle so that

*The methods developed for drag prediction are given in Appendix B and the data accumulated on the atmosphere are given in Appendix A.

F. H. Clauser
PREPARED BY: R. E. Krueger   DOUGLAS AIRCRAFT COMPANY, INC.   PAGE: 113
DATE: May 2, 1946   SANTA MONICA   PLANT   MODEL: #1033
TITLE: PRELIMINARY DESIGN OF SATELLITE VEHICLE   REPORT NO. SM-11827

FORM 25-8-1
(REV. 8-45)

Chapter 9

it reasonably well follows the desired trajectory and is established within permissible limits of error on the chosen orbit. The present chapter concerns itself with the first topic. The second will be the subject of the following chapter.

In our present work, we shall find that it will be necessary to apply forces of control normal to the flight path to obtain desirable trajectories. These forces cannot be applied without incurring losses. Consequently it will be necessary to anticipate the methods of control proposed in the following chapter in order that the losses incurred by this control may properly be included in the calculations. This method of control consists of movable vanes in the rocket jet stream by means of which the entire vehicle is rotated so that a component of the thrust is applied in any direction normal to the flight path. If T is the rocket thrust and α is the angle between flight path and rocket jet axis (for brevity, α will be referred to as the "tilt") then the control force produced is $T \sin \alpha$ and the effective thrust along the flight path is reduced by $T(1-\cos\alpha)$. If α can be kept less than 15°, the reduction in effective thrust is less than 4%. When the desired orbital altitude was 100 miles, it was found that the tilt could be kept within this limit. However, to achieve an altitude of 300 miles, tilts of greater than 15° were required and a revised scheme was found to be necessary as will be seen presently.

In considering possible trajectories, we see that air resistance, starting from its initial value of zero, will first rise rapidly as the speed increases, then less rapidly as altitudes of reduced density are reached. We shall see later that, for the vehicles we have considered, the drag reaches a maximum at altitudes of about 10,000 ft. to 20,000 ft.

PREPARED BY: F. H. Clauser
E. W. Krueger
DOUGLAS AIRCRAFT COMPANY, INC.
PAGE: 114

DATE: May 2, 1946   SANTA MONICA   PLANT   MODEL: #1033

TITLE: PRELIMINARY DESIGN OF SATELLITE VEHICLE   REPORT NO. SM-11827

Chapter 9

Beyond this the decrease in density reduces the drag faster than the corresponding increase in speed. By the time 150,000 ft. has been reached the drag has become a factor of minor importance. It is apparent that the initial portion of the trajectory should be nearly vertical so as to reduce as much as possible the portion of the flight path affected by drag.

First let us consider the following trajectory. The vehicle is launched nearly vertically. As it accelerates upward, gravity will curve the trajectory toward the horizontal in the direction of the initial inclination*. If this initial condition of launching is correctly selected, it is seen that the trajectory will be sufficiently curved without the application of control forces so that the vehicle is on a circular orbit at the end of powered flight as shown by trajectory A in the accompanying figure. It is clear that only one such initial condition of launching exists for a given program of acceleration. If the initial launching is too nearly vertical, the trajectory will

end up at a higher altitude, inclined out into space as indicated by B. If insufficiently near vertical, it will end up at a lower altitude, inclined toward the earth as indicated by C. We are forced to conclude that if our vehicle characteristics are already chosen (i.e. weights, thrusts, etc.) there is only one altitude for the orbit into which it can be launched by this method. Initial calculations showed that for the vehicles

*Mathematically, the initial point of launching is a complicated singularity. Mechanically, this means that a short set of guiding rails will have to be provided.

FORM 25-8-1
(REV. 8-43)

PREPARED BY: F. H. Clauser
R. W. Krueger   DOUGLAS AIRCRAFT COMPANY, INC.   PAGE: 115

DATE: May 2, 1946   SANTA MONICA   PLANT   MODEL: #1033.

TITLE: PRELIMINARY DESIGN OF SATELLITE VEHICLE   REPORT NO. SM-11827

Chapter 9

considered in this report, this altitude was about 35 miles, considerably short of our two successive goals of 100 miles and 300 miles.

We are next faced with the question of how we should apply control forces so that the trajectory* will end up on a circular orbit of greater altitude. Of the three trajectories shown in the figure above, B offers the most obvious possibilities. If downward forces are applied during the late portions of B, it is conceivable that this trajectory can be curved sufficiently to end tangent to a circular orbit. A rigorous examination of the equations of motion shows that this is the best way of obtaining the desired increase in altitude. One might be inclined to question this result on the ground that an application of downward control forces is inefficient when attempting to gain more altitude. Actually, the control forces have little direct effect on the altitude, which is gained primarily by the increased steepness of the earlier portions of the trajectory. The control forces serve primarily to insure a horizontal tangent at the trajectory's end point.

By the use of tilt, it was found possible to determine satisfactory trajectories for orbits at altitudes of 100 miles without exceeding 15°, angle of tilt. However, when the desired altitude was increased to 300 miles, the tilt angles became so large that the losses in effective thrust were excessive. In seeking means of avoiding these losses it was found that **the judicious insertion of an extended period of coasting** in the thrust program would accomplish the desired result. A little study showed that this coasting could be most effectively used if it came late -

*For convenience, we shall refer to that portion of the path traversed before the end of powered flight simply as the trajectory. After powered flight, the path will be called the orbit.

FORM 25-2-1
(REV. 5-42)

PREPARED BY: F. H. Clauser
R. W. Krueger

DOUGLAS AIRCRAFT COMPANY, INC.

PAGE: 116

DATE: May 2, 1946

SANTA MONICA PLANT

MODEL: #1033

TITLE: PRELIMINARY DESIGN OF SATELLITE VEHICLE

REPORT NO. SM-11827

Chapter 9

in the thrust program. In order to avoid the necessity of shutting down a rocket motor and firing it up again, the coasting period was always inserted in the interval between the discarding of the next to the last stage and the firing of the last stage. Unfortunately, insufficient time was available to pursue this study as far as was desirable; however the tentative conclusions indicate that optimum conditions for coasting correspond to a long, slightly inclined coasting trajectory during which altitude is gained surprisingly slowly, followed by a final stage of rocket power during which very little tilt is used.

The equations governing the motion of the vehicle during its acceleration along the trajectory are derived in Appendix C. They are

$$\frac{dV}{dt} = - g \left(1 - \frac{2h}{R}\right) \sin \theta + \frac{T}{m} \cos \alpha - \frac{D}{m} ,$$

$$\frac{d\theta}{dt} = \frac{V}{R} \left(1 - \frac{h}{R}\right) \cos \theta + 2\Omega - g \frac{\left(1 - \frac{2h}{R}\right)\cos \theta}{V} + \frac{T \sin \alpha}{mV}$$

where V is the velocity along the
flight path,

T is the rocket thrust,

D is the drag,

m is the mass of the vehicle,

h is the altitude above the
earth,

R is the radius of the earth,

Ω is the angular velocity of
the earth,

t is the time,

g is the acceleration of gravity,

θ and α are angles explained in the figure

PREPARED BY: F. H. Clauser
R. W. Krueger   DOUGLAS AIRCRAFT COMPANY, INC.   PAGE: 117

FORM 28-5-1
(REV. 8-45)

DATE: May 2, 1946   SANTA MONICA   PLANT   MODEL: #1033

TITLE: PRELIMINARY DESIGN OF SATELLITE VEHICLE   REPORT NO. SM-11827

Chapter 9

It is impossible to obtain explicit analytic solutions of these equations
for the cases we are considering. Instead we resorted to a step by step
method of solution in which the intervals were chosen with sufficient
care to insure that an accuracy of better than 1% was maintained in the
final values. A detailed set of sample calculations is give in Appendix D.

When the desired altitude was 100 miles, the calculations were
carried out in the following manner:

Four Stage Alcohol-Oxygen Rocket-No Coasting - The calculations were made
for each of three ratios of fuel weight to gross weight so that by inter-
polation, the amount of fuel necessary to attain the correct final
velocity could be predicted. The trajectory for the first half(in time) of
the first stage is taken as a vertical path. At this point a constant
angle of tilt is applied and this is carried through to the end of the
second stage. This calculation was carried out for each of three fixed
angles of tilt, so that the results could be interpolated for any inter-
mediate tilt. In the meantime, an independent set of calculations had
been proceeding in which the equations were worked backwards, beginning
at the end of the last stage, with the vehicle on the orbit, and computing
the reverse history along the trajectory back to the beginning of the
third stage, where these calculations were connected up with those proceeding the other way. These reverse calculations were also carried out
for three fixed angles of tilt. When all these calculations were complete,
cross plots of trajectory inclination and altitude at the junction point
were made. From these plots, values of tilts for both sets of calculations
could be selected so that the juncture was continuous for both altitude
and inclination. It will be remembered that each of these sets of calcula-

CRM 25.5.1
(REV. 5.43)
PREPARED BY: F. H. Clauser
R. W. Krueger   DOUGLAS AIRCRAFT COMPANY, INC.   PAGE: 118
DATE: May 2, 1946 (Corr, 5-24-46) SANTA MONICA _____ PLANT   MODEL: #1033
TITLE: PRELIMINARY DESIGN OF SATELLITE VEHICLE   REPORT NO. SM-11827

Chapter 9

tions had been made for a series of fuel weight ratios. The final results were cross plotted so that the velocity at the juncture was also continuous.

In Figures 1 and 2 are shown the flight characteristics and trajectory for our proposed design of an alcohol-oxygen rocket. It will be noticed, that for the particular weight ratios used in this design, the final altitude was 165 miles. For this altitude, the tilt required during the last two burning periods was 35°, which was so large that significant losses in effective thrust occured.

Two Stage Hydrogen-Oxygen Rocket - No Coasting - The method used for the hydrogen-oxygen rocket was substantially the same as that described above. However, when it became apparent from the alcohol-oxygen rocket results that it would be impossible to reach altitudes of 300 miles without the use of excessive tilt, further effort along these lines was discontinued. Instead, attention was concentrated on the use of coasting as a more efficient means of obtaining altitude. The calculations were revised as follows:

Four Stage Alcohol-Oxygen Rocket - Coasting - Instead of the juncture occurring at the end of the second stage, it was now placed at the beginning of the fourth stage. A constant angle of tilt was maintained from the middle of the first stage to the end of the third stage. This was followed by a variable amount of coasting. A set of calculations, working backward from the end of the last stage, was made with several fixed values of tilt. A sufficient number of values of all parameters

FORM 25 BP

Analysis PRELIMINARY DESIGN OF SATELLITE VEHICLE

Prepared by L. G. Krüger

Date May 2, 1946   L. E. Clauser

DOUGLAS AIRCRAFT COMPANY, INC.

Page 119

Model 1103

Report No. 1827

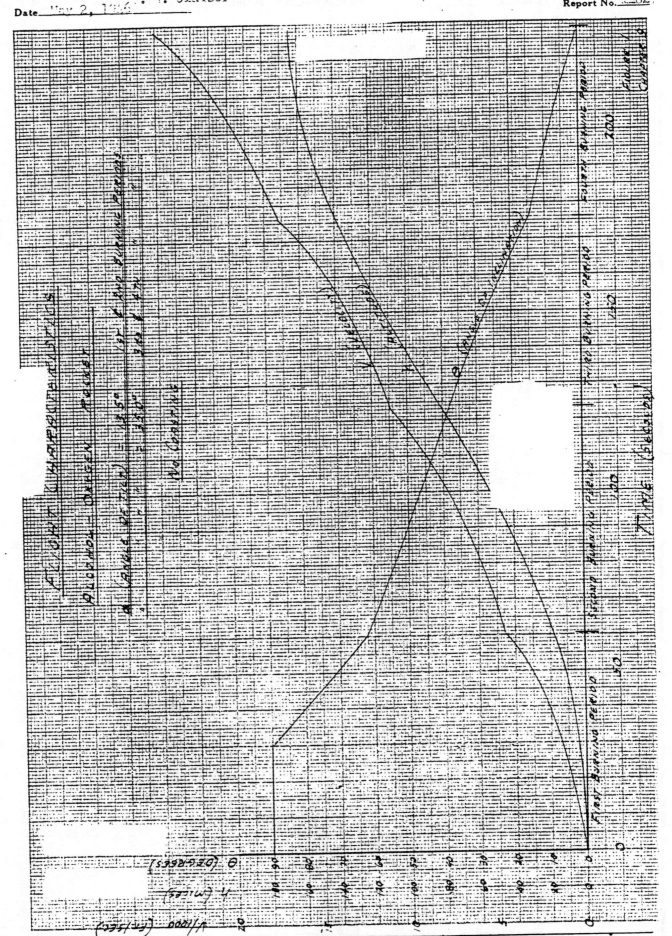

FORM 25 BP

Analysis PRELIMINARY DESIGN OF SATELLITE VEHICLE   Page 120
Prepared by _____   DOUGLAS AIRCRAFT COMPANY, INC.   Model 1033
Date _____   Report No. 1122

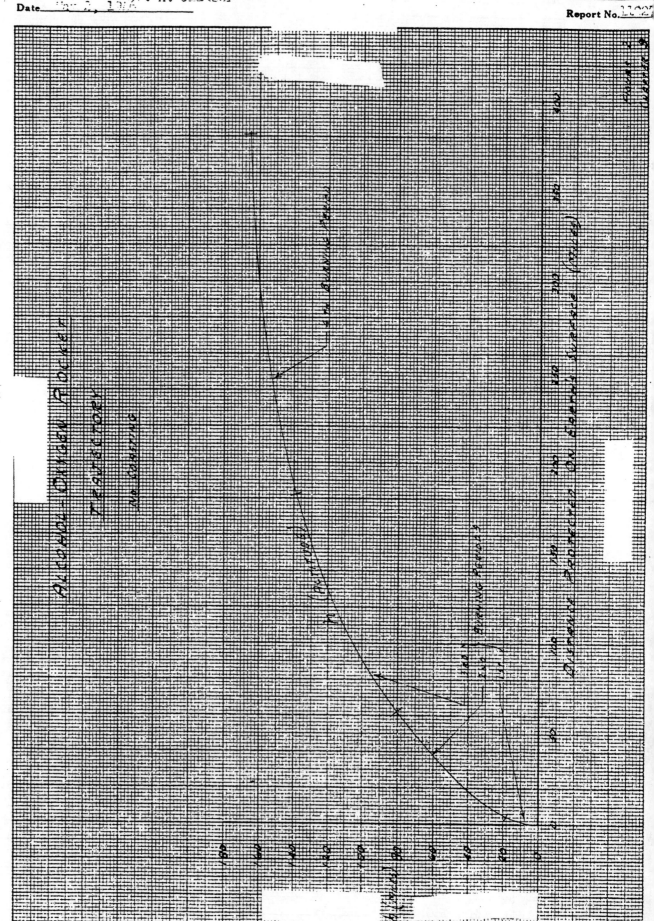

PREPARED BY: F. H. Clauser. R. W. Krueger   DOUGLAS AIRCRAFT COMPANY, INC.   PAGE: 121

DATE: May 2, 1946   SANTA MONICA   PLANT   MODEL: #1033

TITLE: PRELIMINARY DESIGN OF SATELLITE VEHICLE   REPORT NO. SM-11827

Chapter 9

was used so that a continuous juncture could be made at the beginning of the fourth stage. Since a variable amount of coasting has been added to the other variations possible, the choice of values to affect a smooth juncture is not unique. Although time was not available for an exhaustive investigation, it is believed that the optimum trajectory is that discussed a few paragraphs above. The results of these calculations for our proposed alcohol-oxygen rocket are shown in figures 3 and 4. It will be seen that the introduction of coasting has increased the altitude to approximately 480 miles. The greatest angle of tilt required was only 13.5°.

Two Stage Hydrogen-Oxygen Rocket.- For this case, the coasting was inserted between the two stages. In other details, it was the same as the alcohol-oxygen rocket. The results are shown in figures 5 and 6. It will be seen that for the particular weight ratios chosen, the fuel was insufficient to give an altitude greater than 150 miles even with the greater efficiency obtainable from coasting. To achieve an altitude of 400 miles, it would have been necessary to approximately double the weight of the vehicle. The reason for this is not that the altitude has a large effect on performance, but that, with two stages, the hydrogen-oxygen rocket is so far from being an optimum design that the gross weight is highly sensitive to changes in performance requirements. A three stage vehicle would have shown substantially superior performance and weight figures.

RM 25 BS
(EV. 7-42)

Analysis PRELIMINARY DESIGN OF SATELLITE VEHICLE

Prepared by R. W. Krueger

Date May 2, 1946   F. H. Clauser

DOUGLAS AIRCRAFT COMPANY, INC.

Page 122

Model #1033

Plant   Report No. 11827

FLIGHT CHARACTERISTICS

FORM 25 BS
(REV. 7-42)

Analysis PRELIMINARY DESIGN OF SATELLITE VEHICLE

Prepared by R. W. Krueger

Date May 2, 1946   F. H. Clauser

DOUGLAS AIRCRAFT COMPANY, INC.

Page 123

Model #1033

Plant

Report No. 11527

ALCOHOL-OXYGEN ROCKET

TRAJECTORY

FIGURE 4

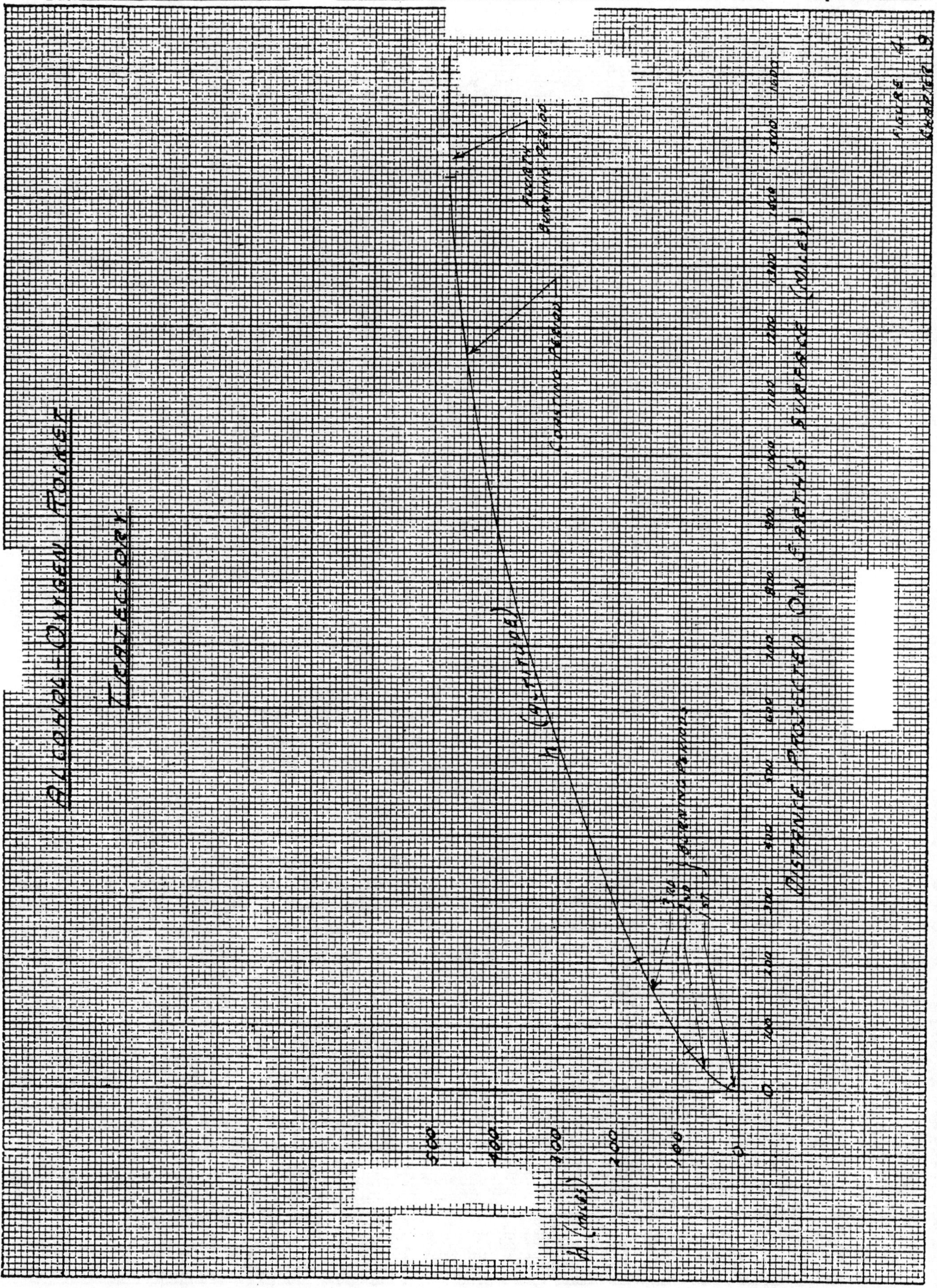

Analysis PRELIMINARY DESIGN OF SATELLITE VEHICLE

Prepared by R. ... Turner   ... Clauser

Date May 2, 1946

DOUGLAS AIRCRAFT COMPANY, INC.

Page 44

Model ....

Report No. 11827

FLIGHT CHARACTERISTICS

HYDROGEN-OXYGEN VEHICLE

Ab COASTING

TIME (SECONDS)

FORM 25 BS
(REV. 7-42)

Analysis PRELIMINARY DESIGN OF SATELLITE VEHICLE
Prepared by R. W. Krueger
Date May 2, 1946

DOUGLAS AIRCRAFT COMPANY, INC.

F. H. Clauser   Santa Monica   Plant

Page 125
Model #1033
Report No. 11827

HYDROGEN + OXYGEN ROCKET

TRAJECTORY

h (ALTITUDE)

3RD BURNING PERIOD

COASTING PERIOD

COASTING PERIOD

1ST BURNING PERIOD

DISTANCE PROJECTED ON EARTH'S SURFACE (MILES)

h (MILES)

Chapter 10

10. METHOD OF GUIDING VEHICLE ON TRAJECTORY

Up to this point, our analysis has considered the design of a vehicle and the selection of its trajectory without regard to the means of guidance to insure that the vehicle follows the prescribed trajectory. In the following paragraphs attention will be devoted to this guidance problem.

In the latter three quarters of the trajectory, the density of the air is so low that in spite of the very great speeds, the dynamic pressures are incapable of adequately guiding the vehicle. Consequently, we are led to the conclusion that we must use reaction motors to obtain forces for guidance.

Two means of obtaining such forces are at once apparent. In the first, the vehicle is rotated (e.g. by means of vanes in the main rocket stream) so that a component of the main rocket thrust is applied in the desired direction. In the second, a small auxiliary rocket oriented normal to the axis of the vehicle is used to obtained the desired guidance forces (several such rockets would have to be provided for control in all directions).

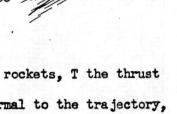

If c is the exhaust velocity available from rockets, T the thrust along the trajectory and L the guidance force normal to the trajectory, then in the first case

$$T = c \frac{dm}{dt} \cos\alpha,$$

Chapter 10

$$L = c \frac{dm}{dt} \sin \alpha,$$

where $\frac{dm}{dt}$ is the mass ejection rate of the main rocket and α is the angle between the trajectory and the vehicle axis. Eliminating α, we have

$$\frac{c \frac{dm}{dt}}{T} = \frac{\frac{dm}{dt}}{\left(\frac{dm}{dt}\right)_0} = \sqrt{1 + \left(\frac{L}{T}\right)^2},$$

where $\left(\frac{dm}{dt}\right)_0$ is the rocket fuel consumption required to produce the thrust if the guidance force were not present.

For the second case, when a small auxiliary rocket is used,

$$T = c \frac{dm}{dt}1,$$

$$L = c \frac{dm}{dt}2,$$

and

$$\frac{\frac{dm}{dt}}{\left(\frac{dm}{dt}\right)_0} = 1 + \frac{L}{T}$$

where $\frac{dm}{dt}1$, $\frac{dm}{dt}2$ and $\frac{dm}{dt}$ are the consumptions of the main rocket, the auxiliary rocket and the total.

In the adjoining figure, the ratio of the consumptions with and without guidance force have been plotted against the ratio of guidance force to thrust. It is at once apparent that case 1 is markedly superior to case 2. In fact with case 1, substantial guidance forces may be obtained without appreciable penalty in thrust. Case 1 will be the method of guidance considered in what follows.

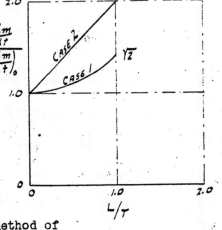

FORM 25-5-1
(REV. 6-45)

PREPARED BY: F. H. Clauser   DOUGLAS AIRCRAFT COMPANY, INC.   PAGE: 128

DATE: May 2, 1946   SANTA MONICA   PLANT   MODEL: #1033

TITLE: PRELIMINARY DESIGN OF SATELLITE VEHICLE   REPORT NO. SM-11827

Chapter 10

If the orbit is not entered with precision it becomes necessary to apply corrections to flight path angle and velocity after the starting trajectory has been completed.

It can easily be demonstrated that, if a thrust is applied normal to the flight path in order to correct the angle, then

$$\frac{\Delta W_{fuel}}{W} = \left(\frac{V}{c}\right) \theta,$$

where

ΔW_{fuel} = fuel weight required,

W = gross weight,

V = flight velocity,

c = exhaust velocity,

θ = change in angle measured in radians.

$\frac{V}{c}$ is approximately equal to 3, so for a one degree correction of angle a weight of fuel equal to 5% of the gross weight is required.

Similarly it can be shown that if resultant velocity is to be corrected, then

$$\frac{\Delta W_{fuel}}{W} = \frac{\Delta V}{c}$$

where ΔV = the velocity increment. A 1% change in resultant velocity requires a fuel weight equal to 3% of the gross weight.

Since these corrections optimistically assume normal rocket efficiency for short period operation, it is evidently very costly in fuel to make corrections of any magnitude.

As an alternative to applying corrections assume that an eccentric orbit can be tolerated, if the eccentricity can be kept within certain

Chapter 10

specified limits. It is then necessary to know the relationships existing between the various orbital parameters. These parameters are as follows:

V_o = the correct velocity for a circular orbit at the starting altitude,

ΔV = the velocity increment above V_o,

β_o = the allowable variation from horizontal (measured positively downward in radians) of the path at the starting altitude,

R_o = the starting radius from the center of the earth,

Δh = the allowable drop in altitude from the starting point,

ΔH_{max} = the maximum minus the minimum distance of the orbit from the earth's surface.

Fig. 1 gives the relationships between Δh, ΔH_{max}, β_o and $\dfrac{\Delta V}{V_o}$. The curves have been plotted from exact equations but for small deviations in angle and velocity the following approximate relations can be used:

$$\beta_o \doteq \sqrt{\left(\frac{\Delta h}{R_o}\right)\left[\left(\frac{\Delta h}{R_o}\right) + 4\left(\frac{\Delta V}{V_o}\right)\right]},$$

or

$$\left(\frac{\Delta h}{R_o}\right) \doteq -\frac{2\Delta V}{V_o} + \sqrt{4\left(\frac{\Delta V}{V_o}\right)^2 + \left(\beta_o\right)^2},$$

or

$$\left(\frac{\Delta H_{max}}{R_o}\right) \doteq 2\sqrt{4\left(\frac{\Delta V}{V_o}\right)^2 + \left(\beta_o\right)^2},$$

or

$$\left(\frac{\Delta H_{max}}{R_o}\right) \doteq 2\left[\left(\frac{\Delta h}{R_o}\right) + 2\left(\frac{\Delta V}{V_o}\right)\right].$$

Assume that the altitude of the vehicle should not drop more that 50 miles below the design starting altitude to be attained at the end

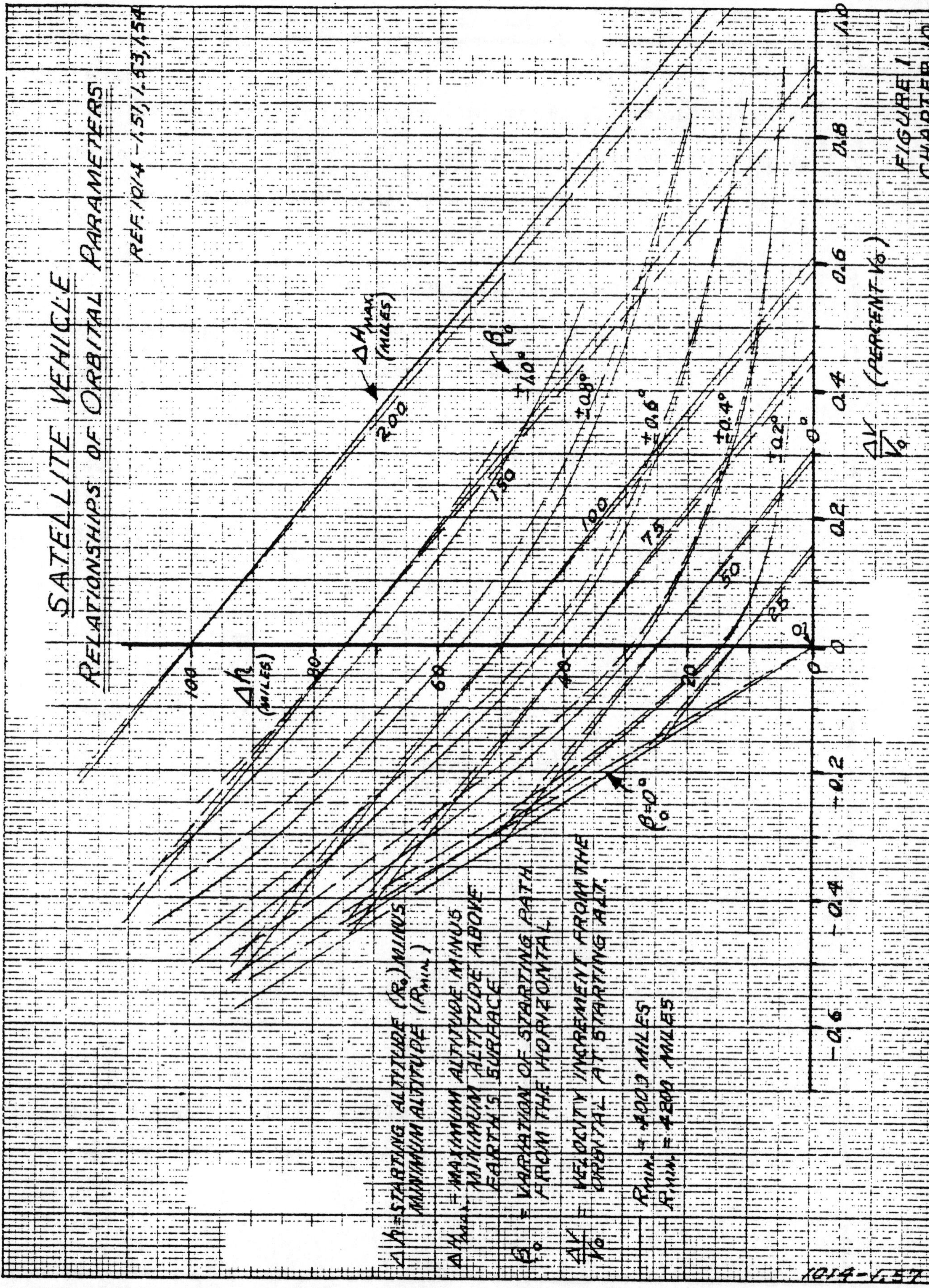

FORM 25 89
REV 7.42

Analysis SATELLITE VEHICLE
Prepared by C. V. STURDEVANT
Date 5-3-46

DOUGLAS AIRCRAFT COMPANY, INC.
SANTA MONICA Plant

Page 130
Model #1033
Report No. SM 11827

SATELLITE VEHICLE
RELATIONSHIPS OF ORBITAL PARAMETERS

REF: 1014-1.51, 1.53, 1.54

FIGURE 1
CHAPTER 10

$\frac{\Delta V}{V_0}$ (PERCENT V_0)

ΔH_{MAX} (MILES)

Δh (MILES)

Δh = STARTING ALTITUDE (R_0) MINUS MINIMUM ALTITUDE (R_{MIN})

ΔH_{MAX} = MAXIMUM ALTITUDE MINUS MINIMUM ALTITUDE ABOVE EARTH'S SURFACE

θ_0 = VARIATION OF STARTING PATH FROM THE HORIZONTAL

$\frac{\Delta V}{V_0}$ = VELOCITY INCREMENT FROM THE ORBITAL AT STARTING ALT.

$R_{MIN} = 4000$ MILES
$R_{MIN} = 4200$ MILES

1014-1.57

FORM 25-S-1
(REV. 8-45)

PREPARED BY: E. W. Graham   DOUGLAS AIRCRAFT COMPANY, INC.   PAGE: 131

DATE: May 2, 1946   Santa Monica   PLANT   MODEL: #1033

TITLE: Preliminary Design of Satellite Vehicle   REPORT NO. SM-11827

Chapter 10

of the starting trajectory, and that the total variation of altitude should not be more than 100 miles. Then if the starting angle is exactly correct ($\beta_o = 0$) the starting velocity may drop 0.3% below the correct value or rise 0.6% above it. However, if the angle can only be established to within $\pm$ 1/2 degree the velocity must fall within the range -0.15% to + 0.40%. The above limits are somewhat arbitrary but illustrate the orders of magnitude involved.

The evident necessity for maintaining close tolerances on the starting conditions for the orbit leads to a preliminary investigation of stability and control requirements.

In the latter part of the starting trajectory a direct control of flight path angle is desirable since this angle must be maintained with extreme accuracy. This requires a restoring moment proportional to deviation of flight direction from the horizontal. It is anticipated that this deviation can be measured by means of a radar equipped ground station which measures both range and angle, computes rate of change of altitude and sends a corresponding control impulse to the vehicle. A beacon will be used in the vehicle to act as a "transponder" and can also be used to convey information from the vehicle to the ground.

In addition to the restoring moment proportional to deviation of flight direction from the horizontal it is necessary to apply either a damping moment depending on flight direction or a restoring moment depending on pitch. The latter would probably be simpler to use. To correct for an unknown eccentric thrust (so that the vehicle will not only be stable but also approach the exact flight angle desired) it is

Chapter 10

also necessary to apply a restoring moment proportional to the integral of flight direction over a period of time. This corresponds actually to an altitude control. Use of integral terms in control problems has been discussed by Weiss* and such a control is used on constant speed propellers.

From stability considerations it can be shown that the restoring moment depending on pitch (or damping moment depending on flight direction, whichever is used) should be large. However, further investigation is required to determine desirable magnitudes for the other terms.

Velocity control can probably be obtained by using an integrating accelerometer which operates a fuel cut-off valve. Such an accelerometer was used on the V-2 with accuracy of 1/2 % over a 60 second period, and it is believed that this accuracy can be improved. At present radar techniques involving radar ranging or Doppler effect do not appear to offer adequate accuracy.

In the early stages of the starting trajectory it should be sufficient to control pitch as a predetermined function of time. For stability, restoring and damping moments can be applied as functions of the difference between actual and desired pitch angles (determined with the aid of a pre-set gyro). In order to approach the desired trim condition a moment should also be applied as a function of the integral of deviation in pitch angle over a period of time, and it may also be necessary to apply a predetermined moment as a function of time to compensate for the calculated curvature of the trajectory.

* "Dynamics of Constant-Speed Propellers" Herbert K. Weiss, Journal of the Aeronautical Sciences, Vol. 10, No. 2, Feb. 1943.

Chapter 10

The areas of the control surfaces in the jet have been briefly studied. The control moments required depend upon the angular accelerations required and upon the fixed disturbing moment, such as that due to a displaced thrust line, The control surface areas shown on the drawings are quite arbitrary but serve to show, in conjunction with figure 2 that design of sufficiently powerful surfaces should not be difficult. This figure shows the variation of control surface area with displacement of the thrust line from the center of gravity in inches for each of the four stages. Actually the maximum error is expected to result from a rotation of the thrust line about the throat of the engine by about 0.5 deg. The resultant displacement is indicated in the figure as the maximum probable displacement. Also shown are the areas required to produce a pitching or yawing acceleration of 10 deg./sec^2. It is seen that the areas shown in the drawings are adequate to overcome the moment due to the maximum probable thrust line displacement and to produce in addition a 10 deg./sec^2 acceleration if a 10 deg. control surface angle is used. Since these areas are far from excessive from a mechanical standpoint, they could be increased if more thorough control studies showed the desirability of obtaining higher angular accelerations.

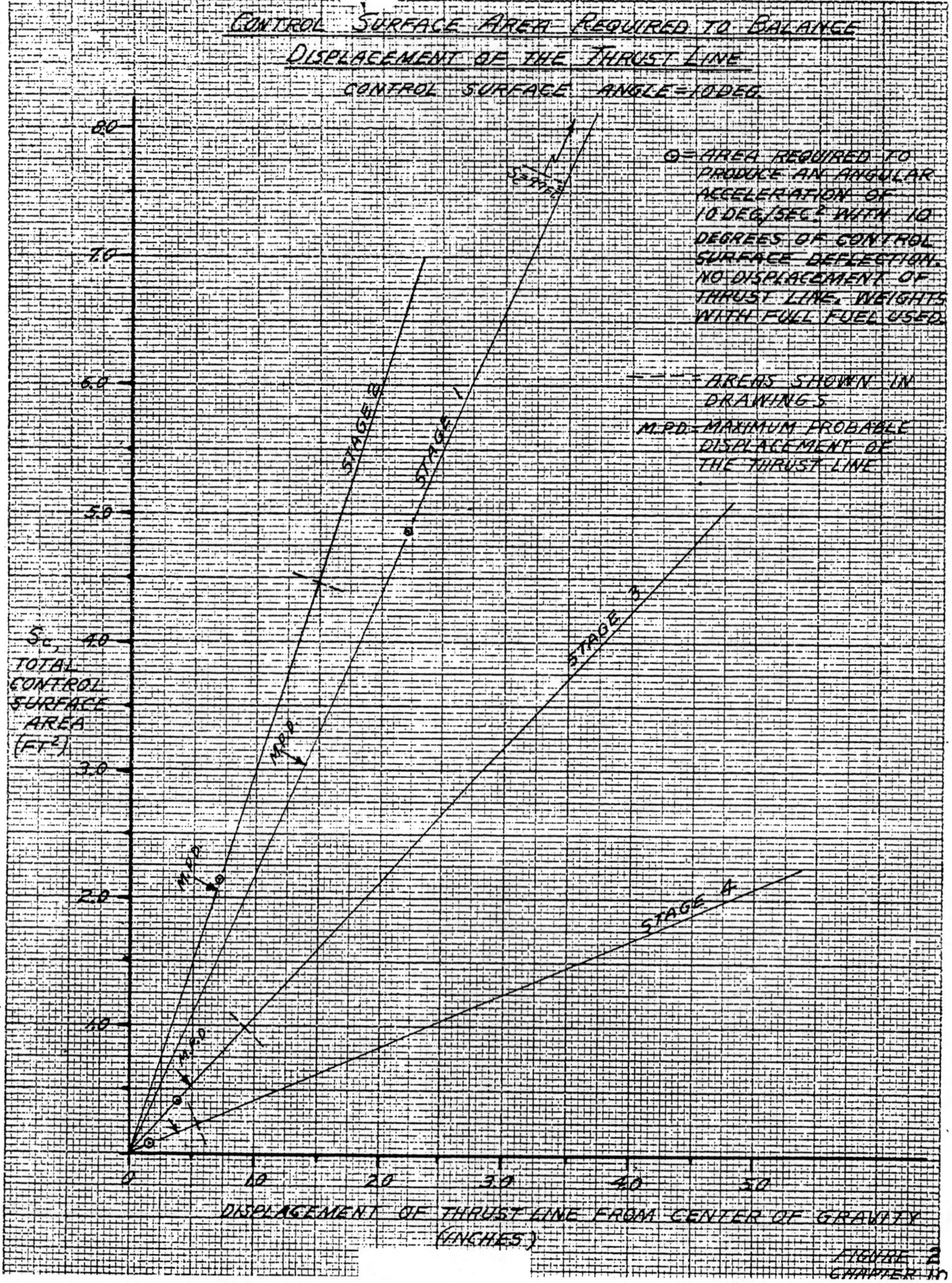

ORM 2565
(REV. 7-45)

Analysis CONTROL
Prepared by R.S. SHEVELL
Date 5-7-46

DOUGLAS AIRCRAFT COMPANY, INC.

S.M.   Plant

Page 134
Model # 1033
Report No. SM 11827

CONTROL SURFACE AREA REQUIRED TO BALANCE
DISPLACEMENT OF THE THRUST LINE
CONTROL SURFACE ANGLE = 10 DEG.

Θ = AREA REQUIRED TO
PRODUCE AN ANGULAR
ACCELERATION OF
10 DEG/SEC² WITH 10
DEGREES OF CONTROL
SURFACE DEFLECTION.
NO DISPLACEMENT OF
THRUST LINE. WEIGHTS
WITH FULL FUEL USED.

----- AREAS SHOWN IN
DRAWINGS
M.P.D. - MAXIMUM PROBABLE
DISPLACEMENT OF
THE THRUST LINE

S_c,
TOTAL
CONTROL
SURFACE
AREA
(FT²)

DISPLACEMENT OF THRUST LINE FROM CENTER OF GRAVITY
(INCHES)

FIGURE 2
CHAPTER III

Chapter 10

The preceding investigations are preliminary only, but suggest the following:

1. It is desirable to rotate the entire vehicle to obtain a component of the jet thrust for purposes of control.

2. Corrections on the orbit are undesirable after it has once been established

3. At the end of the starting trajectory the flight path angle should be accurate to $\pm \frac{1}{2}$ degree and the velocity should be accurate to $\pm 0.3\%$.

4. This accuracy and the necessary stability can probably be attained.

Chapter 11

11. PROBLEMS AFTER ORBIT IS ESTABLISHED

Once the vehicle has been established in its orbit at the desired altitude various other problems arise in connection with the satisfactory operation of the vehicle. For example, the vehicle will be constantly exposed to the possibility of being hit by meteorites of all sizes and some of which will be travelling at very high speed. Also, at such high altitudes the intense heating of the vehicle by the sun is a problem to be considered. Radio contact must be maintained. These and other problems connected with the satisfactory operation of the vehicle are discussed below.

Meteorites. (The Probability of a Meteorite Hitting a Satellite Vehicle Traveling in the Upper Atmosphere.) It is well known that a great many meteorites enter the earth's atmosphere each day. If a body should be situated in the upper atmosphere at altitudes where meteorites are observed with high frequency, the question arises as to what are the chances that the body will be struck by a meteorite and if a strike does occur what are the probabilities that the meteorite will seriously damage or otherwise interfere with the motion of the body.

Meteorites* are discrete masses of matter from outer space which enter the earth's atmosphere. Judging from those which are large enough to survive the journey through the air and reach the ground, and which are then called fallen meteorites, they are composed mainly

---

* Leonard, F. C.: Meteorites: Immigrants From Space. Publications of the Astronomical Society of the Pacific, Vol. 57, No. 334; p. 1, Feb. 1945.

Chapter 11

of stony matter (similar to igneous rock) and metallic nickeliferous iron. Like fallen meteorites, the relative amounts of iron and stony matter in the meteorites may be expected to vary greatly ranging from almost all iron to almost all stone. However, it is quite likely that the stony meteorites are more prevalent than iron meteorites by a factor of more than ten, although a stony meteorite itself may contain some 25 percent iron by mass. It will be assumed that the meteorites consist mainly of stony matter.

Meteorites vary greatly in size ranging from something smaller than a pin head or grain of sand up to the large meteoritic masses found on the earth which weigh 10 or 20 tons or more. (According to Leonard, Reference 1, meteorites may be of any magnitude whatever, from the size of the tiniest solid particle to that of a mass of planetary dimensions, and are the smallest discrete astronomical bodies. The term meteor is properly used to denote the luminous phenomenon which results from the motion of a meteorite through the earth's atmosphere.) It is estimated that the weight of the average fallen meteorite is 220 pounds before entering the atmosphere and that this is reduced to about 44 pounds by the time the earth's surface is reached. However, meteorites which are large enough to reach the earth's surface occur with such low frequency, 5 or 6 a day for the whole earth, that they need not be considered here.

Chapter 11

Typical values of velocity and altitude as determined by observations of certain bright meteors are given in Table 1 which is taken from Hoffmeister.[2]

As one might expect, it is seen from Table I that the higher the meteorite makes its appearance the greater is its velocity. Velocities ranging from 80,000 to 250,000 ft. per sec. are quite common. According to Watson[3] most meteors appear at a height of about 300,000 feet regardless of their brightness and may be taken to have an average atmospheric velocity of about 150,000 feet per second. Thus, at an altitude of 500,000 feet, where the body is assumed to be situated, most all of the meteorites will be intact and will not have suffered complete dissipation. At this altitude the body will therefore be exposed to practically all of the meteorites which enter the atmosphere.

The number, size and mass of meteorites entering the atmosphere each day is given in Table 2 which is based on a table given by Watson (ibid., p. 115).

The visual magnitude of a meteor is expressed in terms of a scale in which numerically large magnitudes represent faint bodies. Two meteors which differ by five magnitudes have a hundred-fold difference in brightness and, since the brightness is directly proportional to

(2) Hoffmeister, C.: Die Meteore. Probleme Der Kosmischen Physik, Band XVII; p. 71, 1937.

(3) Watson, F. G.: Between the Planets. The Blakiston Company, Philadelphia; p. 93, 1945.

Chapter 11

TABLE 1

VELOCITY AND ALTITUDE OF BRIGHT METEORS

(From Hoffmeister, Ref. 2)

Velocity		Mean Height of Appearance			Mean Height of Disappearance		
$\frac{km}{sec.}$	$\frac{ft.}{sec.}$	km	ft.	Number	km	ft.	Number
10	32,810	66	216,000	4	28	91,900	4
20	65,600	92	302,000	35	44	144,200	34
30	98,500	112	368,000	93	47	154,000	93
40	131,200	131	430,000	107	47	154,000	107
50	164,000	144	472,000	74	47	154,000	75
60	196,800	138	452,500	54	55	180,500	54
70	229,600	149	489,000	31	63	206,500	31
80	262,500	183	600,000	11	74	243,000	12
90	295,000	197	645,000	15	98	321,500	15
100	328,100	217	712,000	3	96	315,000	3
110	361,000	226	741,000	6	136	446,000	6

1 km. = 3,281 ft.

Chapter 11

TABLE 2

THE NUMBER, MASS, AND SIZE OF METEORITES ENTERING THE ATMOSPHERE EACH DAY

(Based on Watson,)

Visual Magnitude	Observed Number	True Number N	Mass Grams	Weight lbs. w	Diameter of Equivalent Sphere, Ft.* d
-3	28,000	28,000	4.0	8.72×10^{-3}	$.427 \times 10^{-1}$
-2	71,000	71,000	1.6	3.53×10^{-3}	$.317 \times 10^{-1}$
-1	180,000	180,000	.630	1.39×10^{-3}	$.232 \times 10^{-1}$
0	450,000	450,000	.250	5.51×10^{-4}	$.1705 \times 10^{-1}$
1	1,100,000	1,100,000	.100	2.20×10^{-4}	1.257×10^{-2}
2	2,800,000	2,800,000	.040	8.72×10^{-5}	$.922 \times 10^{-2}$
3	6,400,000	7,100,000	.016	3.53×10^{-5}	$.683 \times 10^{-2}$
4	9,000,000	18,000,000	.0063	1.39×10^{-5}	$.500 \times 10^{-2}$
5	3,600,000	45,000,000	.0025	5.51×10^{-6}	$.367 \times 10^{-2}$
6		110×10^{6}	.0010	2.20×10^{-6}	$.2705 \times 10^{-2}$
7		280×10^{6}	.00040	8.72×10^{-7}	$.1986 \times 10^{-2}$
8		710×10^{6}	.00016	3.53×10^{-7}	1.471×10^{-3}
9		18×10^{8}	.000063	1.39×10^{-7}	1.078×10^{-3}
10		45×10^{8}	.000025	5.51×10^{-8}	$.793 \times 10^{-3}$
15		45×10^{10}	2.5×10^{-7}	5.51×10^{-10}	1.705×10^{-4}
20		45×10^{12}	2.5×10^{-9}	5.51×10^{-12}	$.367 \times 10^{-4}$
25		45×10^{14}	2.5×10^{-11}	5.51×10^{-14}	$.793 \times 10^{-5}$
30		45×10^{16}	2.5×10^{-13}	5.51×10^{-16}	1.705×10^{-6}

lbs. = grams $\times 2.205 \times 10^{-3}$

*Based on a specific gravity of 3.4.

Chapter 11

the mass, they represent a hundred-fold difference in mass. A meteor just visible to the naked eye has a magnitude of 5 while the full moon has a magnitude of -14. Also, it will be noticed from Table 2 that when the magnitude differs by five units, the number of meteors changes by a factor of 100. In this way the table may be extended to include smaller and smaller meteorites (numerically larger magnitudes).

However, there is a limiting magnitude beyond which there can be few, if any, meteorites, and according to Watson, p. 116, this limiting size is a meteorite of magnitude 30. This is explained by the fact that for a particle smaller than this the solar radiation pressure is sufficient to repel any particle to such an extent that it could not remain in the solar system.

In Table 2, figures are included for meteorites down to the smallest possible size, magnitude 30. The sizes have been computed on the basis that the meteorite is a sphere composed mainly of stony matter which according to Whipple[4], has a specific gravity of 3.4. The variation of size with magnitude is presented in Fig. 1.

It is seen that a great range in size and mass is represented in the table and the question immediately arises, especially for the very small particles, as to what velocities are to be associated with the various sizes. If it is assumed that the meteorites move in parabolic orbits at the same distance from the sun as the earth, a meteorite

(4) Whipple, F. L.: Meteors and the Earth's Upper Atmosphere. Reviews of Modern Physics, Vol. 15, No. 4; p. 252, Oct. 1943.

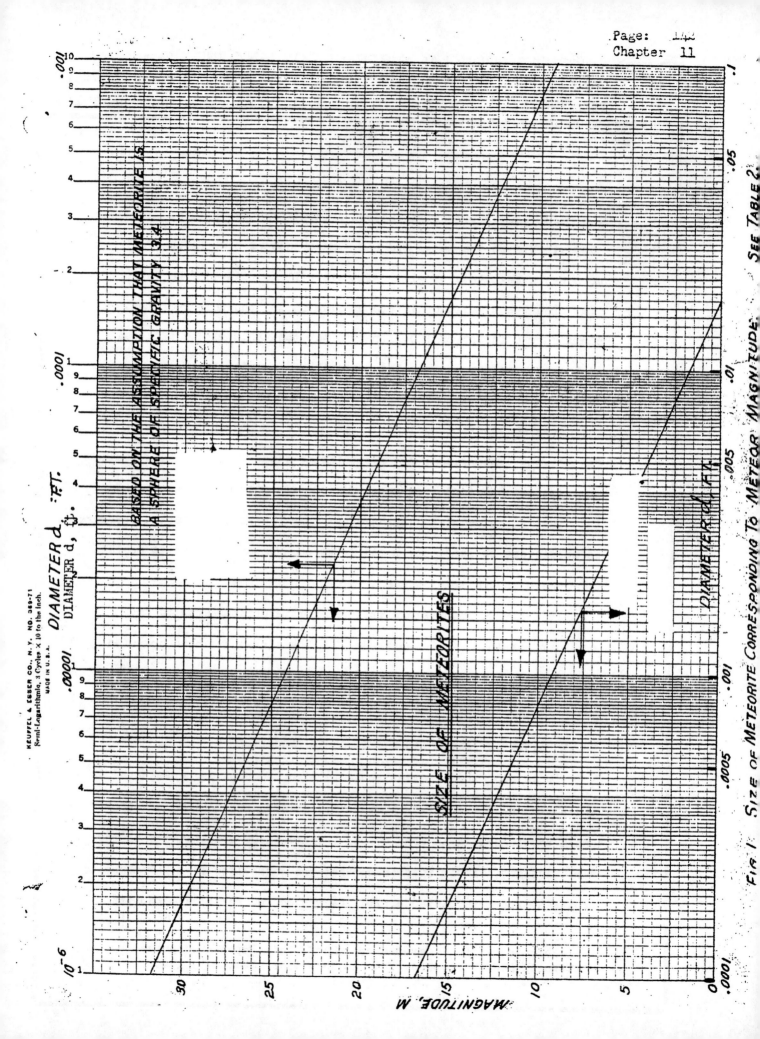

FIG. 1. SIZE OF METEORITE CORRESPONDING TO METEOR MAGNITUDE. SEE TABLE 2.

Chapter 11

encountering the earth head-on will enter the outer atmosphere with a
speed of 250,000 ft. per sec., while one overtaking the earth will
enter the atmosphere with a speed of only 43,300 ft. per sec. These
speeds would be the same for meteorites of all sizes since there has
not yet been any deceleration resulting from air resistance. However,
once the meteorite has entered the atmosphere, deceleration must take
place and this must certainly be greater for a small meteorite than
for a larger one, assuming that both enter the atmosphere with the
same speed. Whipple, loc. cit. p. 252 gives as a universally accepted
expression for the deceleration,

$$\frac{dV}{dt} = \frac{-\gamma a}{m^{1/3}} \rho V^2, \text{--------------------------(1)}$$

where V = velocity,

m = mass,

ρ = density of the atmosphere,

$a(m)^{2/3}$ = effective cross sectional area,

γ = non-dimensional form factor depending on the shape of the
meteorite but independent of velocity.

For a sphere, $\gamma = 1/3$ and if the density is 3.4, $a = 0.53$.

The rigorous investigation of the deceleration would involve a
study of the variations of γ, a, ρ, and especially m, as a function
of time or distance. Since time is not available to carry on such a
study at the present and since at this stage of the project, approximate values will be satisfactory, we therefore adopt the following
approximate method.

Chapter 11

Assume that γ, a, and m are independent of the motion of the meteorite through the air and since values for density and its variation in the very high atmosphere are lacking, replace ρ by an average or effective value $\bar{\rho}$.

The differential equation may then by written

$$\frac{d\,(V^2)}{V^2} = \frac{2\gamma a\bar{\rho}}{m^{1/3}}\,dh$$

and therefore

$$V_f = V_i\,(e)^{-\dfrac{\gamma\,a\bar{\rho}\,\Delta h}{m^{1/3}}} = (e)^{-\dfrac{k\,\Delta h}{m^{1/3}}} \quad \cdots\cdots\cdots(2)$$

In this expression V_i is the velocity of the meteorite when it enters the atmosphere, and V_f its velocity after it has fallen a vertical distance Δh through the air. As a maximum condition it will be assumed that the meteorite enters the atmosphere head-on so that its speed is given by $V_i = 47.4$ mi./sec. $= 250,000$ ft./sec. See p. 9 of Leonard, loc. cit.

From Table 1 it is estimated that bright meteors (about magnitude 2) have a velocity of about 200,000 ft./sec. at 100 miles altitude (528,000 ft.) so that $V_f = 200,000$ ft./sec. Although the height of the approximate upper limit of the atmosphere is not known, auroral

---

* At this stage of the analysis, the density values derived in Appendix A were not yet available. However, it is believed that a more exact treatment of the density variation with altitude would not appreciably change the results derived here by the approximate method.

Chapter 11

observations indicate the pressure of atmosphere at 1000 km. (622 miles) and this figure will be used to represent the altitude of the effective limit of the atmosphere. Since we are interested in the meteorite velocity at 100 miles altitude,

$$\Delta h = 522 \text{ miles} = 840 \text{ km} = 8.40 \times 10^7 \text{ cm.}$$

For a meteor of magnitude 2 it is found from Table 2 that $m_2 = .040$ gram, and hence $m_2^{1/3} = .343$. The constant k may now be evaluated giving $k = 0.9 \times 10^{-9}$.

It was found in Table 2 that a change of one magnitude corresponded to a change in mass by a factor of 2.5. Thus if all masses are referred to that of a meteor of magnitude 2 one may write

$$m_M = \frac{m_2}{[2.5]^{M-2}} = \frac{.040}{[2.5]^{M-2}} , \quad \text{- - - - - - - - - - -(3)}$$

and Eq. (2) may then be written

$$V_M = 2.5 \times 10^5 \, [e]^{-.22 \times [2.5]^{\frac{M-2}{3}}} \quad \text{ft. per sec., - - - (4)}$$

at 100 miles altitude. This function is plotted in Fig. 2 where it is seen that for magnitudes greater than 5, the velocity decreases rapidly.

Since the vehicle may also operate at altitudes higher than 100 miles, say up to 400 miles, it becomes necessary to obtain an

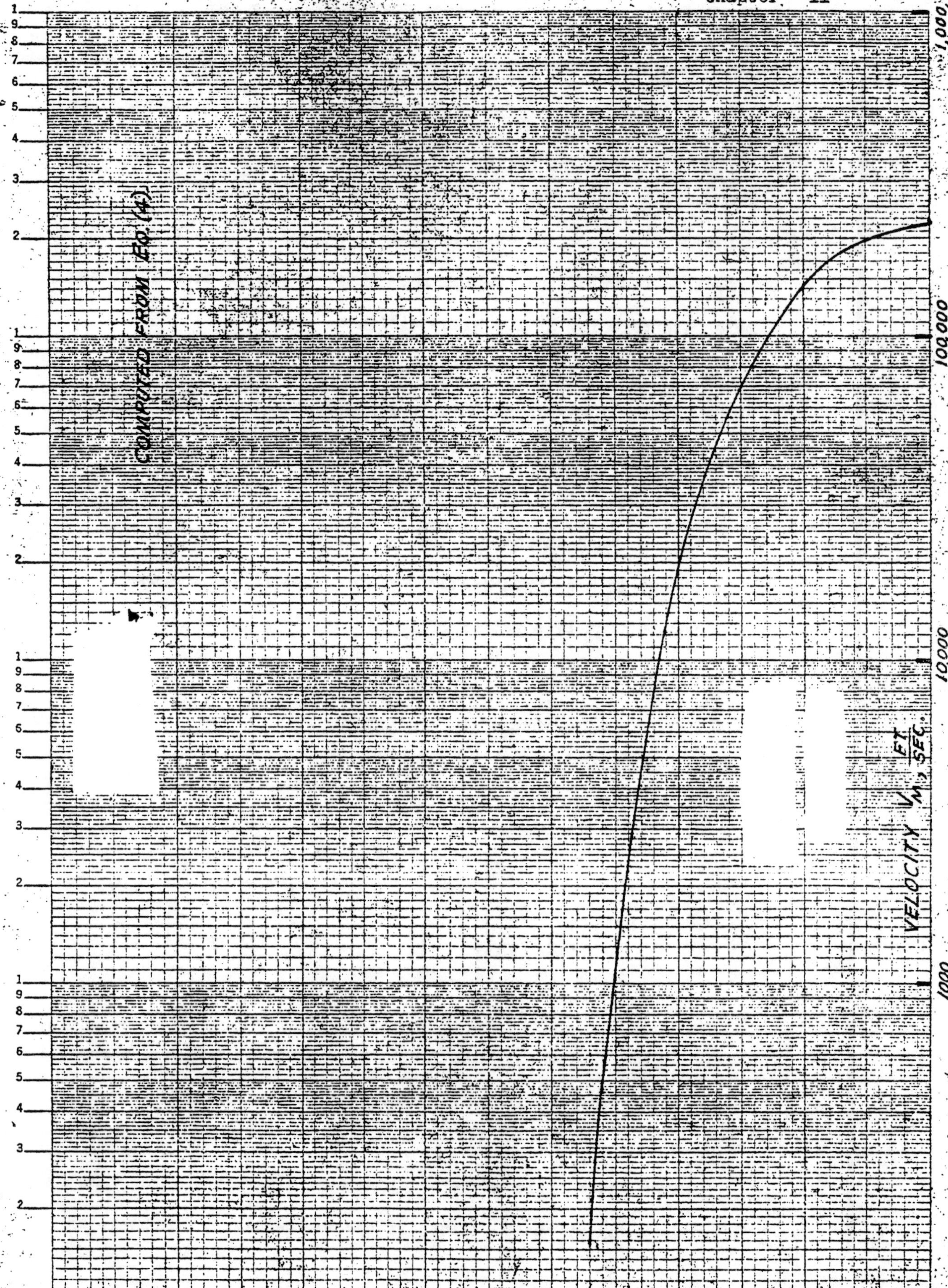

COMPUTED FROM EQ. (4)

VELOCITY V_M, $\frac{FT.}{SEC.}$

FIG. 2   ESTIMATED METEORITE VELOCITY AT 100 MILES ALTITUDE.

Chapter 11

estimate of V_M for this upper altitude limit also. At an altitude
as high as 400 miles it is not possible to obtain any estimate of
the deceleration from Table 1. Furthermore, the effective density
value $\bar{\rho}$ from 400 to 622 miles is certainly much different than that
from 100 to 622 miles and therefore the value found above for the
deceleration factor k would not apply to this much higher altitude.
In fact at such a high altitude it is not entirely unlikely that
the deceleration would be negligible. We thus have the two extremes
within which the velocity must lie, that of no change in velocity,
and that with velocity given by using the value $k = 0.9 \times 10^{-9}$
found above for the 100 mile altitude.

The velocity corresponding to this latter limit is given by
the equation

$$V_M = 2.5 \times 10^5 \left[e \right]^{-.094 \ \times \left[2.5 \right]^{\frac{M-2}{3}}} , \quad - - - - - - - (4A)$$

corresponding to $\Delta h = 622 - 400 = 222$ miles.

These two extremes are shown in Fig. 2A where the solid curve
which has been drawn in to represent a compromise between the two
extremes will be used as an estimate of the velocities prevailing
at 400 miles altitude. It is seen that at these extremely high altitudes, the smaller meteorites maintain a fairly high velocity.

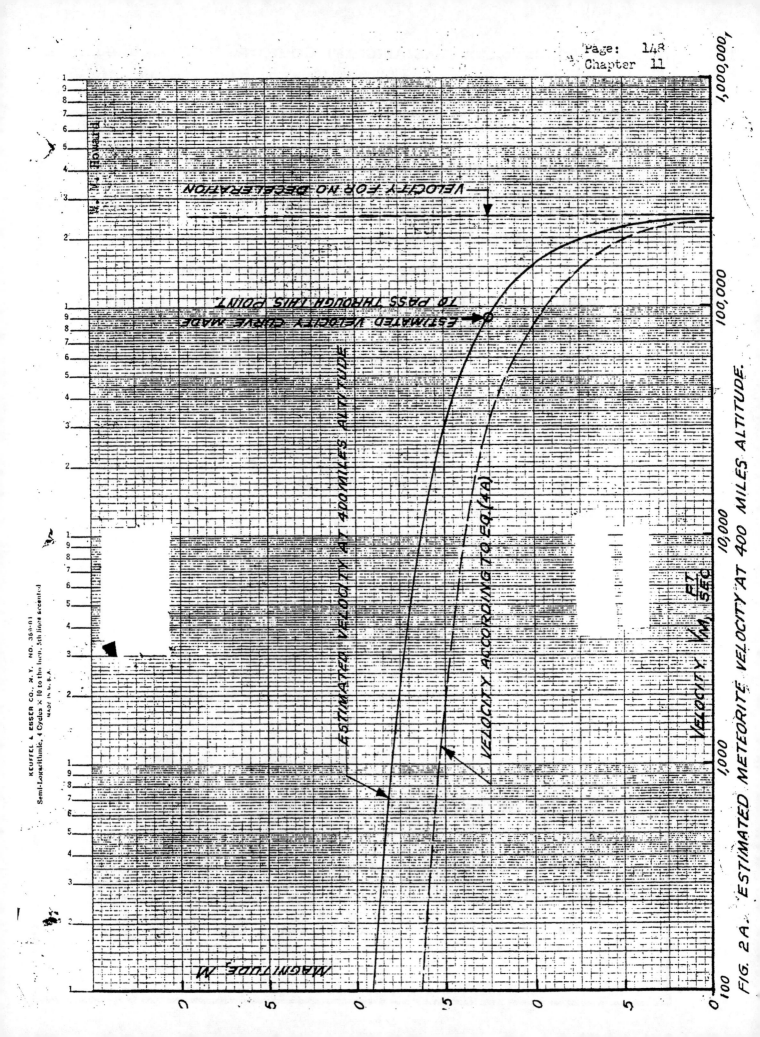

FIG. 2A. ESTIMATED METEORITE VELOCITY AT 400 MILES ALTITUDE.

Chapter 11

Having the relation between $\bar{V}_M$ and M given by Figs. 2 and 2A it is now necessary to determine what is the smallest meteorite which will penetrate through (perforate) the skin of the satellite vehicle. There seems to be very little, if any, information available, either theoretical or experimental, on the penetration of metal plate by very small but extremely high speed particles. Bethe, [5] has apparently worked on this problem to some extent but unfortunately his paper is not available.

However, the indications are that in the case of normal impact of a small but very high speed particle on a metal plate in which the speed of the particle is large compared to the velocity of propagation of plastic deformation in the plate, the particle penetrates as though the plate were perfectly deformable like a fluid. In this case the differential equation for the motion of the particle through the plate is

$$m\frac{dv}{dt} = \frac{m}{2}\frac{d(v^2)}{ds} = -C_D\frac{\rho_0}{2}\pi\frac{d^2}{4}v^2, \quad - - - - - - - - - (5)$$

where

   m = mass of particle = $\frac{4}{3}\pi\frac{d^3}{8}\rho_M$ (assumed spherical),

   d = diameter of particle,

   s = distance of penetration into the plate,

   ρ_0 = density of the plate = $2.8\ \frac{gm}{cm^3}$ for dural,

(5): Bethe, H. A.: "Attempt at a Theory of Armor Penetration", Unnumbered Report of the Frankfort Arsenal, 1941.

Chapter 11

ρ_M = density of particle = $3.4 \frac{gram}{cm^3}$ for a meteorite,

$\sigma = \frac{\rho_o}{\rho_M} = .825$,

C_D = drag coefficient = $2/3$, see Epstein.   (6)

The equation becomes

$$\frac{d(V^2)}{V^2} = -\sigma \frac{ds}{d},$$

which when integrated gives

$$s = \frac{d}{\sigma} \log \frac{V_M^2}{V^2},$$

where V_M is the velocity of the particle as it first strikes the plate and V is its velocity after it has penetrated a distance s. When the speed of the particle has dropped to about 5 times the plastic deformation velocity V_1, it will be assumed that this law of penetration is no longer to be used. In the range in which it is to be used the equation is then written

$$s = 2 \frac{d}{\sigma} \log \frac{V_M}{5V_1} \text{ - (6)}$$

When the speed of the particle is less than 5 times the plastic deformation speed it will be assumed that the penetration takes place according to one of the armor penetration formulas.

(6): Epstein, P. S.: Proc. Nat. Acad. Sci., Vol 17, p. 532, 1931.

Chapter 11

(7)

The well-known DeMarre armor penetration formula for plain wrought iron is

$$t^{.65} = \frac{.333 \times 10^{-4} \, w^{\frac{1}{2}} \, v}{d^{.75}} , \quad \text{-------------(7)}$$

where

t = penetration in ft.,

d = diameter of particle in ft.,

w = weight of particle in pounds,

v = velocity of particle in ft./sec.

In this equation, V is the velocity necessary to perforate a thickness t of wrought iron by a particle of weight w and diameter d.

(7a)

The Watertown Arsenal uses the formula

$$t = F \frac{mV^2}{d^2} , \quad \text{--------------------(8)}$$

where

t = thickness penetrated,

m = mass of projectile,

F = Thompson coefficient,

d = diameter of projectile,

V = velocity of projectile.

(8)

The work of Duwez and Clark on penetration of copper by high

(7): The United States Naval Institute: Naval Ordnance, Annapolis, Md. 1939, p. 301.
(7a): Sullivan, J.F.: An Empirical Approach To The Efficient Design of Armor For Aircraft. Watertown Arsenal Laboratory, Watertown, Mass. Experimental, Report No. WAL 710/506, Jan., 1944, p. 9.
(8): NDRC Report Number M-317

speed small caliber bullets directly verifies the velocity squared relation for speeds of the order of 4000 ft./sec. Their results can be expressed by

$$\frac{t}{d} = 5.9 \times 10^{-7} \, v^2. \quad \text{- - - - - - - - - - - - - - - - - - -} \quad (9)$$

Thus the experimental data give rise to penetration formulas proportional to $v^{1.5}$ or v^2. Since the results of Duwez and Clark are considered the best to use in the present study, a formula corresponding to (9), which refers to copper, will be used in the ballistic range. Since the ultimate strengths of dural and copper are of the same order of magnitude and since the ballistic penetration is closely connected with this property of the metal, it will be assumed that the results for copper also apply to dural within the degree of approximation of the equations.

Eq. (9) was obtained from experiments with 0.224 inch diameter projectiles. If it is assumed that the energy loss per unit penetration is proportional to the frontal area A of the projectile and also constant over the ballistic range (as shown in ref. (8)) it follows that

$$\frac{\triangle KE}{t} = \frac{\frac{1}{2} \, mV^2}{t} = CA,$$

where $\triangle KE$ = change in kinetic energy, and

        C = a constant

Chapter 11

In the tests, the projectile weight was 69 grains and was brought to rest in about 2 inches of copper. This gives

$$\frac{1}{2} mV^2 = 13.8 \times 10^3 \quad \frac{\text{ft.-lb.}}{\text{ft.}}$$

and

$$C = \frac{\frac{1}{2} mV^2}{tA} = 5.03 \times 10^7 \quad \frac{\text{ft.-lb.}}{\text{ft.}^3}.$$

The corresponding penetration formula for a sphere would be

$$\frac{C \pi d^2}{4} = \frac{\frac{1}{2} \rho_M \frac{\pi}{6} d^3 V^2}{t} \text{, or}$$

$$\frac{t}{d} = \frac{\rho_M}{3C} V^2.$$

Taking $\rho_M = 3.4 \frac{\text{gram}}{\text{cm}^3} = 6.6 \frac{\text{slug}}{\text{ft.}^3}$ and $C = 5.03 \times 10^7$

as determined from the firing tests on copper, the penetration formula for a sphere becomes

$$\frac{t}{d} = 4.4 \times 10^{-8} V^2 . \text{ - - - - - - - - - - - - - - - - (9A)}$$

The velocity of plastic deformation in compression is known to be around 1000 ft. per sec. so that the limiting speed used in the [9] fluid-flow equation (6) will be taken to be 5,000 ft. per sec.

(9) NDRC Report Number M-302.

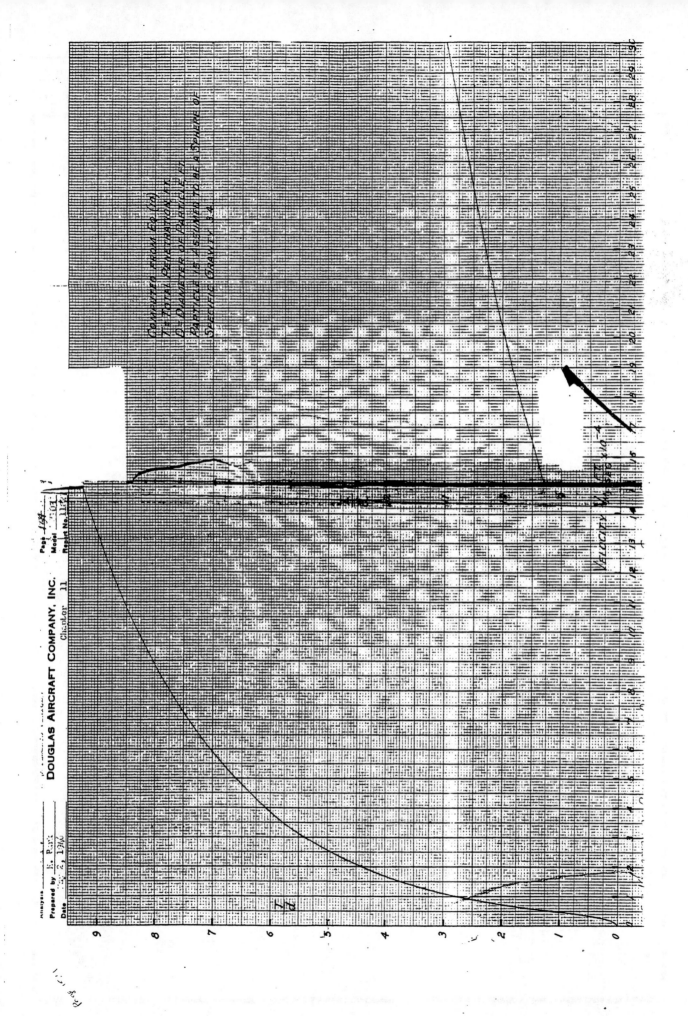

Computed from Eq. (a)
T = Total Penetration in ft.
D = Diameter of particle, ft.
Particle is assumed to be a sphere of
specific gravity 3.4

VELOCITY, Vo, ft./sec. ×10⁻²

$\frac{T}{d}$

Chapter 11

This corresponds to a plastic deformation Mach number of 5. Thus to compute the total penetration for a particle (meteorite) having a speed greater than 5000 ft. per sec., Eq. (6) is first used to compute the penetration s at which the speed is slowed down to 5,000 ft. per sec. The remainder of the penetration t is then computed from Eq. (9) using V = 5,000 ft. per sec. The total penetration is then given by the sum s + t. Letting T=s+t, the computation can be simplified by joining Eqs. (6) and (9) at V_M = 5,000 ft. per sec. and the total penetration T is then given by

$$\frac{T}{d} = \begin{cases} 4.4 \times 10^{-8} \, V_M^{\,2}, & \text{for } 0 \leq V_M \leq 5,000 \text{ ft./sec.} \\ \\ 1.1 + 2.4 \log_e \dfrac{V_M}{5,000}, & \text{for } V_M \geq 5,000 \text{ ft./sec.} \end{cases} \quad \text{-------(10)}$$

The variation of $\frac{T}{d}$ with V_M is shown in Fig. 3.

These equations are essentially empirical and neglect effect of shape of projectile and influence of the physical properties of the metal on the critical velocities. Since the basis for the formulas lies in extrapolating rather meager ballistic data and theories to the small sizes of meteorites, considerable error may be expected. The present results may, however, serve to give an indication of the order of magnitude of the impact effects.

Chapter 11

Using the results contained in Figs. 1 and 2 which connect the diameter and velocity of a meteorite with the magnitude of its meteor, the penetration at an altitude of 100 miles can be expressed directly as a function of meteor magnitude. This is tabulated in Table 3 and presented graphically in Fig. 4.

In a similar fashion, using Fig. 2A, the penetration at 400 miles altitude is obtained as a function of meteor magnitude, and this is presented in Table 3A and Fig. 4A.

From Tables 3 and 3A or Figs. 4 and 4A, one may see immediately how thick the skin (assumed to be of dural) of the satellite vehicle must be to withstand perforation by meteorites of different sizes (magnitudes). Thus at 100 miles altitude, for a meteor of magnitude 0, the skin, according to the analysis, would have to be at least 2.08 inches thick in order to resist perforation. For a skin thickness of .05 in. = .00416 ft. which represents the order of thickness of metal commonly used in aircraft design, it is seen that the smallest meteorite which will perforate corresponds to about magnitude 9 or 10. For velocities less than about 1000 ft. per sec. the particles would probably not penetrate the plate at all, but simply bounce off. In view of the penetration results presented in Table 3 for the 100 mile altitude, it is seen that, as far as presenting perforation hazard is concerned, meteorites of corresponding magnitude greater than 12 can certainly be completely disregarded.

Chapter 11

TABLE 3

PENETRATION OF DURAL PLATE BY METEORITES - ALTITUDE 100 MILES

Magnitude of Meteor M	Velocity of Meteorite V_M, ft./sec.	Penetration Ratio $\frac{T}{d}$	Diameter of Meteorite d, ft.	Total Penetration Distance T, ft.	Total Penetration Distance T, inches
0	220,000	10.2	.017	.1734	2.08
1	213,000	10.12	.0126	.12751	1.53
2	203,000	10.00	.0092	.09200	1.104
3	189,000	9.83	.0068	.06684	.8021
4	170,000	9.58	.0050	.04790	.5748
5	143,000	9.17	.0037	.03393	.40716
6	112,000	8.58	.0027	.02317	.27804
7	82,500	7.83	.0020	.01566	.18792
8	56,500	6.90	.00147	.01014	.12168
9	36,000	5.84	.00107	.006249	.074988
10	19,500	4.32	.00079	.003413	.040956
11	8,200	2.28	.00058	.001322	.015864
12	2,500	0.25	.00042	.000105	.001260
13	550	0.03	.00032	.0000096	.000115

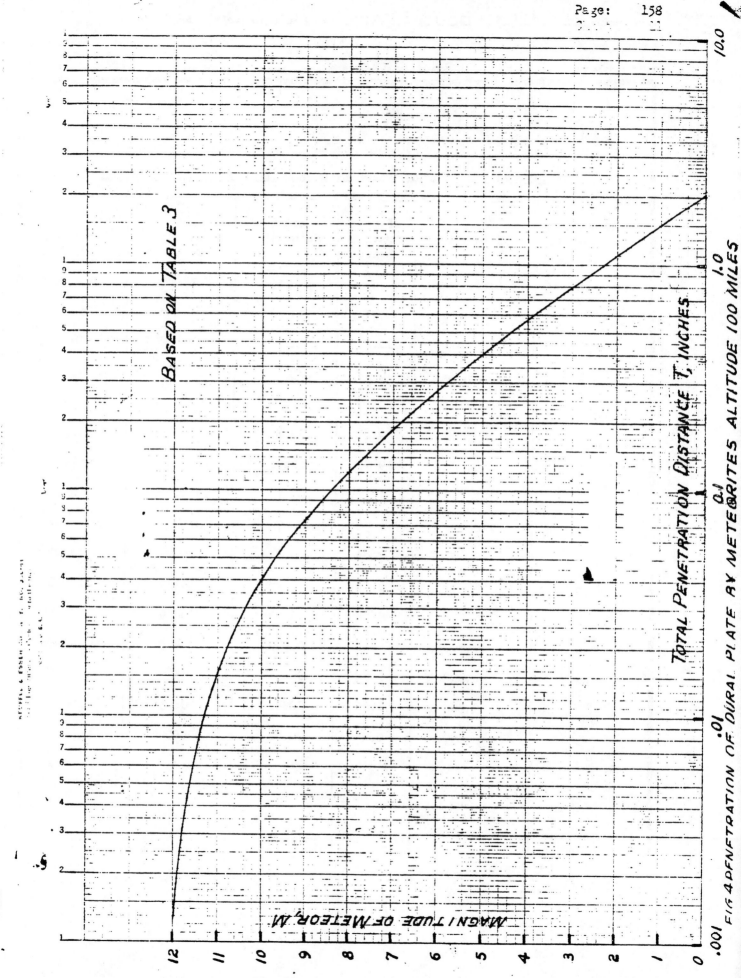

BASED ON TABLE 3

MAGNITUDE OF METEOR, M

TOTAL PENETRATION DISTANCE T, INCHES

FIG. 4 PENETRATION OF DURAL PLATE BY METEORITES ALTITUDE 100 MILES

TABLE 3A

PENETRATION OF DURAL PLATE BY METEORITES - ALTITUDE 400 MILES

MAGNITUDE OF METEOR M	VELOCITY OF METEORITE V_M FT./sec	PENE-TRATION RATIO $\frac{T}{d}$	DIAMETER OF METEORITE d, ft.	TOTAL PENETRATION DISTANCE T, ft.	TOTAL PENETRATION DISTANCE T, inches
0	244,000	10.43	.017	.1775	2.130
1	242,000	10.41	.0126	.1312	1.572
2	240,000	10.39	.0092	.0955	1.145
3	236,500	10.35	.0068	.0704	0.844
4	231,000	10.30	.0050	.0515	0.618
5	225,000	10.24	.0037	.0379	0.454
6	215,000	10.14	.0027	.0274	0.329
7	203,000	10.00	.0020	.0200	0.240
8	190,000	9.84	.00147	.0145	0.174
9	173,000	9.62	.00107	.0103	0.124
10	154,300	9.35	.00079	.0074	0.089
11	131,000	8.96	.00058	.00520	0.0623
12	105,400	8.43	.00042	.00354	0.0424
13	78,000	7.68	.00032	.00246	0.0295
14	52,000	6.70	.00023	.00154	0.0185
15	30,000	5.40	.00017	.00092	0.0110
16	12,450	3.35	.00013	.00046	0.0055
17	4,400	0.90	.00009	.00008	0.00096
18	1,020	0.05	.00007	.000004	0.000048

Table 3A
Chapter 11

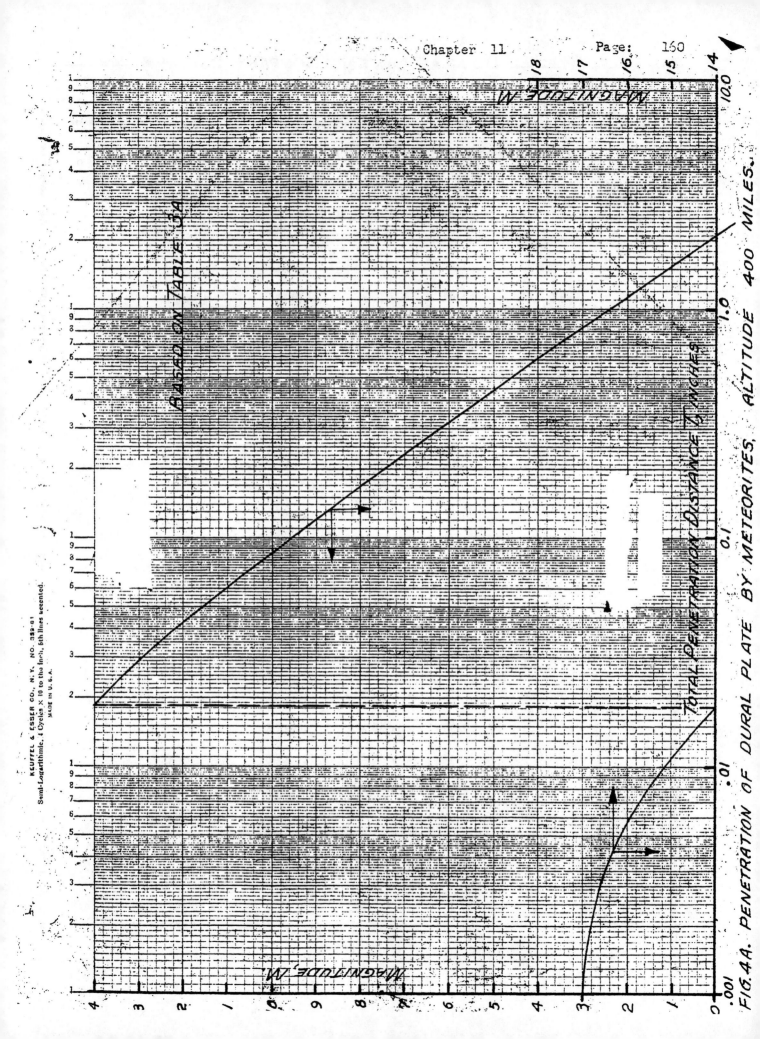

FIG. 4A. PENETRATION OF DURAL PLATE BY METEORITES, ALTITUDE 400 MILES.

Chapter 11

Comparing the values in Tables 3 and 3A it is found that for magnitudes of 5 or less the increase in altitude from 100 to 400 miles does not appreciably affect the penetration. For the smaller sizes the difference in penetration at the two altitudes becomes more marked, although for magnitudes greater than 15, the penetration becomes negligible.

In case the skin is made of material of greater strength and hardness than dural, stainless steel for example, the penetration would be expected to be correspondingly less. In this connection, however, it is worth noting that according to the experiments reported in ref. 8, when the projectile size was much smaller than the plate thickness, dural gave greater resistance to perforation than face hardened steel when the comparison is made on the basis of the weight per unit area of plate.

Aside from the problem of perforation by a meteorite, the question also arises as to what sort of average impact force (averaged over a long interval of time) is to be expected as a result of meteorite hits. For a given magnitude (size) M, if $\bar{n}$ is the average number of hits per hour, W the weight in pounds, V the velocity in ft. per sec., the average impact force $\bar{F}$ is simply

$$\bar{F} = \frac{\bar{n}}{3600} \frac{W}{g} V, \text{ lbs.} \quad - - - - - - - - - - - - - - - (10A)$$

From Table 2 it will be found that the product of $\bar{n}W$ is constant and equal to 1.73×10^{-12}. Thus the average force of impact is given by

$$\bar{F} = 1.5 \times 10^{-17} V, \text{ lbs.},$$

Chapter 11

which shows that even with the highest velocities considered here, this average force would be far too small to in any way affect the performance of the vehicle.

Having arrived at figures for the penetration by the meteorites of different sizes, it now remains to find the probabilities that the vehicle will be struck by these particles. The following notation will be used.

N = number of meteorites (either total or of specified size) entering the earth's atmosphere in each 24 hour period.

A_e = number of square feet of atmospheric surface at a height of 500,000 feet. The number which will be used here is $A_e = 4\pi \times (21.39)^2 \times 10^{12} = 57.4 \times 10^{14}$ sq. ft. Radius of earth = 20.89×10^6 ft.

A_b = planform area of the satellite vehicle, sq. ft. The number which will be used is for a triangular planform 60 ft. x 32 ft. which gives the value, A_b = 960 sq. ft.

p_{1+} = probability that at least one hit will occur in the time T.

p_o = probability that no hit will occur in the time T. $(p_o = 1 - p_{1+})$.

p_1 = probability that exactly one hit will occur in the time T.

$T(0.5)$ = time interval such that the vehicle has a 50 to 50 chance of not being hit.

$T(0.99)$ = time interval such that the vehicle has a 100 to 1 change of not being hit.

$T(0.999)$ = time interval such that the vehicle has a 1000 to 1 chance of not being hit.

TABLE 4

PROBABILITIES OF HIT OF A METEORITE AND SATELLITE VEHICLE
PROBABILITIES AND TIME INTERVALS BASED ON NUMBER OF METEORITES OF ONE SIZE ONLY

	−3	0	2	5	6	7	8	9	10	12	15	20	25	30
Magnitude, M	−3	0	2	5	6	7	8	9	10	12	15	20	25	30
Average number of hits per hour, $\bar{H}$	2.0×10^{-10}	3.14×10^{-9}	2.0×10^{-8}	3.14×10^{-7}	7.9×10^{-7}	2.0×10^{-6}	4.95×10^{-6}	1.25×10^{-5}	3.14×10^{-5}	1.97×10^{-4}	3.14×10^{-3}	$.314$	31.4	3140
Average number of hours between hit	5.0×10^{9}	3.18×10^{8}	5.0×10^{7}	3.18×10^{6}	1.27×10^{6}	5.0×10^{5}	2.02×10^{5}	$80{,}000$	$31{,}800$	5080	318	3.18	$.0318$	$.000318$
P_{1+} for 24 hours	4.8×10^{-9}	7.54×10^{-8}	4.8×10^{-7}	7.54×10^{-6}	1.9×10^{-5}	4.8×10^{-5}	1.19×10^{-4}	3.0×10^{-4}	7.54×10^{-4}	4.73×10^{-3}	$.073$	$1-5.49 \times 10^{-4}$	$1-10^{-326}$	$1-10^{-32{,}600}$
P_0 for 24 hours	$1-4.8 \times 10^{-9}$	$1-7.54 \times 10^{-8}$	$1-4.8 \times 10^{-7}$	$1-7.54 \times 10^{-6}$	$1-1.9 \times 10^{-5}$	$1-4.8 \times 10^{-5}$	$1-1.19 \times 10^{-4}$	$1-3.0 \times 10^{-4}$	$1-7.54 \times 10^{-4}$	$1-4.73 \times 10^{-3}$	9.27×10^{-1}	5.49×10^{-4}	10^{-326}	$10^{-32{,}600}$
P_1 for 24 hours	4.8×10^{-9}	7.54×10^{-8}	4.8×10^{-7}	7.54×10^{-6}	1.90×10^{-5}	4.8×10^{-5}	1.19×10^{-4}	3.0×10^{-4}	7.54×10^{-4}	4.73×10^{-3}	6.98×10^{-2}	4.13×10^{-3}	10^{-326}	$10^{-32{,}600}$
P_{1+} for 120 hours	2.4×10^{-8}	3.77×10^{-7}	2.4×10^{-6}	3.77×10^{-5}	9.5×10^{-5}	2.4×10^{-4}	6.0×10^{-4}	1.5×10^{-3}	3.77×10^{-3}	2.33×10^{-2}	$.313$	$1-2.0 \times 10^{-16}$	$1-10^{-1630}$	$1-10^{-163{,}000}$
P_0 for 120 hours	$1-2.4 \times 10^{-8}$	$1-3.77 \times 10^{-7}$	$1-2.4 \times 10^{-6}$	$1-3.77 \times 10^{-5}$	$1-9.5 \times 10^{-5}$	$1-2.4 \times 10^{-4}$	$1-6.0 \times 10^{-4}$	9.98×10^{-1}	9.96×10^{-1}	9.77×10^{-1}	6.67×10^{-1}	2.0×10^{-16}	10^{-1630}	$10^{-163{,}000}$
P_1 for 120 hours	2.4×10^{-8}	3.77×10^{-7}	2.4×10^{-6}	3.77×10^{-5}	9.5×10^{-5}	2.4×10^{-4}	6.0×10^{-4}	1.5×10^{-3}	3.77×10^{-3}	2.28×10^{-2}	2.59×10^{-1}	2.54×10^{-15}	10^{-1630}	$10^{-163{,}000}$
$T(0.5)$, hours	3.46×10^{9}	2.2×10^{8}	3.46×10^{7}	2.2×10^{6}	8.8×10^{5}	3.46×10^{5}	1.4×10^{5}	$55{,}400$	$22{,}000$	3520	220	2.20	$.0220$	$.00022$
$T(0.99)$, hours	5.0×10^{7}	3.2×10^{6}	5.0×10^{5}	3.2×10^{4}	1.27×10^{4}	5000	2020	800	318	50.7	3.18	$.0318$	$.000318$	3.18×10^{-6}
$T(0.999)$, hours	5.0×10^{6}	3.2×10^{5}	5.0×10^{4}	3.2×10^{3}	1.27×10^{3}	500	202	80	31.8	5.07	$.318$	$.00318$	3.18×10^{-5}	3.18×10^{-7}

Definition of symbols:

P_{1+} = Probability of at least 1 hit

P_0 = Probability of no hit

P_1 = Probability of only one hit

$T(0.5)$ = Time interval to give a 50 to 50 chance of no hit

$T(0.99)$ = Time interval to give a 1.0 to 1 chance of no hit

$T(0.999)$ = Time interval to give a 1000 to 1 chance of no hit

TABLE 5

PROBABILITIES OF HIT OF A METEORITE AND ON SATELLITE VEHICLE

PROBABILITIES AND TIME INTERVALS BASED ON NUMBER OF METEORITES OF A GIVEN SIZE PLUS ALL THOSE OF LARGER SIZE

Magnitude, M	-3	0	2	5	7.5	8	9	10	12	15	20	25	30
Average hits per hour, $\bar{n}$	2.0×10^{-10}	5.07×10^{-9}	3.28×10^{-8}	5.20×10^{-7}	1.1×10^{-6}	8.22×10^{-6}	2.07×10^{-5}	5.20×10^{-5}	3.28×10^{-4}	5.20×10^{-3}	5.20×10^{-1}	5.20×10^{1}	5.20×10
Average number of hours between hits, $\bar{t}$	5.0×10^{9}	1.92×10^{8}	3.05×10^{7}	1.92×10^{6}	7.63×10^{5}	1.22×10^{5}	5.16×10^{4}	1.92×10^{4}	3.05×10^{3}	1.92×10^{2}	1.92	1.92×10^{-2}	1.92×10
P_{1+} for 24 hours	4.8×10^{-9}	1.22×10^{-7}	7.88×10^{-7}	1.28×10^{-5}	7.88×10^{-5}	1.97×10^{-4}	4.97×10^{-4}	1.25×10^{-3}	7.88×10^{-3}	$.117$	$1-2.63\times10^{-5}$	$1-15^{542}$	$1-10^{-54200}$
P_0 for 24 hours	$1-4.6\times10^{-9}$	$1-1.22\times10^{-7}$	$1-7.88\times10^{-7}$	$1-1.25\times10^{-5}$	$1-7.88\times10^{-5}$	$1-1.97\times10^{-4}$	$1-4.97\times10^{-4}$	$1-1.25\times10^{-3}$	$1-7.88\times10^{-3}$	$.883$	2.63×10^{-5}	10^{-542}	10^{-54200}
P_1 for 24 hours	4.8×10^{-9}	1.22×10^{-7}	7.88×10^{-7}	1.25×10^{-5}	7.88×10^{-5}	1.97×10^{-4}	4.97×10^{-4}	1.25×10^{-3}	7.88×10^{-3}	1.10×10^{-1}	3.28×10^{-4}	10^{-540}	10^{-54230}
P_{1+} for 120 hours	2.4×10^{-8}	6.10×10^{-7}	3.94×10^{-6}	6.25×10^{-5}	3.94×10^{-4}	9.85×10^{-4}	2.48×10^{-3}	6.25×10^{-3}	$.038$	$.465$	$1-1.3\times10^{-27}$	$1-10^{-2710}$	$1-10^{-27103}$
P_0 for 120 hours	$1-2.4\times10^{-8}$	$1-6.10\times10^{-7}$	$1-3.94\times10^{-6}$	$1-6.25\times10^{-5}$	$1-3.94\times10^{-4}$	$1-9.85\times10^{-4}$	$1-2.48\times10^{-3}$	$1-6.25\times10^{-3}$	$.962$	$.535$	1.3×10^{-27}	10^{-2710}	10^{-27103}
P_1 for 120 hours	2.4×10^{-8}	6.10×10^{-7}	3.94×10^{-6}	6.25×10^{-5}	3.94×10^{-4}	9.85×10^{-4}	2.48×10^{-3}	6.25×10^{-3}	3.79×10^{-7}	3.34×10^{-1}	8×10^{-26}	10^{-2710}	10^{-27103}
$T(0.5)$, hours	3.46×10^{9}	1.36×10^{8}	2.11×10^{7}	1.33×10^{6}	5.28×10^{5}	8.46×10^{4}	3.58×10^{4}	1.33×10^{4}	2.11×10^{3}	1.33×10^{2}	1.33	1.33×10^{-2}	1.33×10
$T(0.99)$, hours	5.0×10^{7}	1.92×10^{6}	3.05×10^{5}	1.92×10^{4}	7.63×10^{3}	1.22×10^{3}	5.16×10^{2}	1.92×10^{2}	3.05×10	1.92	1.92×10^{-2}	1.92×10^{-4}	1.92×10
$T(0.999)$, hours	5.0×10^{6}	1.92×10^{5}	3.05×10^{4}	1.92×10^{3}	7.63×10^{2}	1.22×10^{2}	5.16×10	1.92×10	3.05	1.92×10^{-1}	1.92×10^{-3}	1.92×10^{-5}	1.5

Definition of symbols:
P_{1+} = Probability of at least 1 hit

P_0 = Probability of no hit

P_1 = Probability of only one hit

$T(0.5)$ = Time interval to give a 50 to 50 chance of no hit

$T(0.99)$ = Time interval to give a 100 to 1 chance of no hit

$T(0.999)$ = Time interval to give a 1000 to 1 chance of no hit

Chapter 11

These probabilities and time intervals are presented in Tables 4 and 5. The formulas by which they are computed are derived in Appendix G.

The meteorites entering the atmosphere are assumed to have a random distribution both as regards their surface distribution over the atmospheric layer surrounding the earth and as regards their occurrence with time. It is assumed that the meteorites travel through the atmosphere along the vertical and that the planform area of the vehicle is normal to the vertical.

The two tables are entirely similar, the only difference being that the values used for N in Table 4 are based on the total number of meteorites of one size only; whereas, the values for N used in Table 5 include the total number of meteorites of a given size plus all those of larger size. At the lower magnitudes these two numbers do not differ appreciably, but at the higher magnitudes, 9 or 10 and higher, the difference is large enough to be considered. See Table 1, Appendix G. For this reason the values in Table 5 are considered to be the more significant and this table will receive the main consideration.

Considering Table 5, it is seen, first of all, that the average time interval between hits does not attain values comparable to the contemplated time of operation of the vehicle (say from 5 to 10 days) until the meteorite size becomes as small as that corresponding to magnitude 14 or 15. Thus, it is seen from the column for $M = 15$, that on the average, the vehicle could operate for 192 hours before

Chapter 11

it would be hit by a meteorite of size corresponding to magnitude 15 or any larger size. For magnitudes greater than 15 the average time between hits becomes relatively small but by this time the meteorites are of such small size and velocity that it does not matter. The probability numbers, of course, show the same general tendencies as do the numbers for $\bar{n}$ and $\bar{t}$.

The most important probability to consider, it would seem, is the probability P_{1+} which is the probability that the vehicle will be hit at least once. Or, stated slightly differently, P_{1+} gives simply the probability that the vehicle will be hit, the number of times it will be hit not being specified. The probability scale is such that a probability of 1 means that the event is certain to occur, while a probability of 0 means the event is certain not to occur. Considering the values of P_{1+} in Table 5 for the 120-hour interval (5 days), for instance, the probability of a hit is less than 1 in a 1000 (i.e. 0.001) for all magnitudes of 8 or less. At magnitude 15, however, the probability has greatly increased and shows that there is only about a 50 - 50 chance that the vehicle will not be hit. Here again, however, the size of the particle becomes so small that even though the probability becomes high, it does not matter as far as damage to the vehicle is concerned. For magnitudes 20 and above, the vehicle is certain to be hit, but it certainly will not matter considering the small size of these particles.

Considering next the probability-based time intervals, we see, for example, that if we specify a 1000 to 1 chance of not being hit by a meteorite of corresponding magnitude 10 or less, the vehicle may operate for 19.2 hours. If the probability number is relaxed down to a 100 to 1 chance of no hit, the operating time increases to 192 hours, etc.

It is interesting to note that at around magnitude 15, the probabilities p_{1+} , p_o and p_1 , all take on comparable values, showing that somewhere in this range of magnitude the occurrence of these three events becomes more or less equally probable.

In general, the probability tables indicate that for the meteorite sizes which are large enough to present a perforation hazard, the probabilities of a hit are quite small, never exceeding about 0.001 (for a reasonable plate thickness say, of 0.10 in.) or about 1 chance in 1000.

Having the relation between T and M (Figs. 4 and 4A) and the relation between p_{1+} and M (Table 5), one may then derive a relation between T and p_{1+}, where p_{1+} is the probability that a meteorite of corresponding magnitude M will just perforate a dural skin thickness of amount T. This relationship has been derived and is shown in Fig. 5, for the two altitudes 100 miles and 400 miles. These curves represent, essentially, the net result of the perforation and probability study when presented in the most usable form.

Since the relation of the type shown in Fig. 5 has been determined only for the case of a 5 day time interval, we shall suppose that the vehicle is to operate for a period of 5 days. We then,

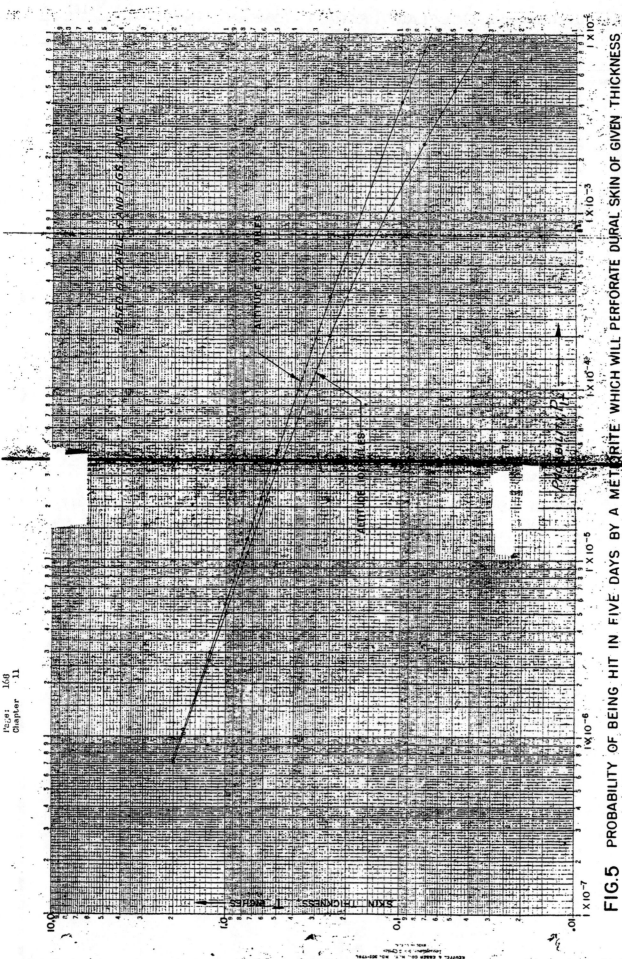

FIG.5 PROBABILITY OF BEING HIT IN FIVE DAYS BY A METEORITE WHICH WILL PERFORATE DURAL SKIN OF GIVEN THICKNESS

Chapter 11

for example, may ask what is the probability that a skin of say 0.12 in. thick dural will be perforated by a meteorite when the altitude is, for instance, 100 miles. Referring to the 100 mile altitude curve of Fig. 5 for $T = 0.12$ in., it is found that $p_{1+} = 1 \times 10^{-3} = .001$. Thus, for the chosen condition the chances are 1000 to 1 that the skin will not be perforated. For a skin thickness $T = .05$ in., the probability is $p_{1+} = .0048$, and in this case the chances are only 208 to 1 that perforation will not occur. There is not much point in considering values of $T < .05$ in. since at least this much thickness would be required simply from considerations of structural strength.

The use of the curves of Fig. 5 may also be considered from the reverse point of view. Assuming operation at 100 miles altitude, suppose we are willing to take a 1000 to 1 chance on the occurrence of perforation, and then ask what the skin thickness must be. For $p_{1+} = .001$ and at 100 miles altitude it is found that $T = 0.12$ in. These examples are sufficient to show how the perforation - probability-time curve is used.

Chapter 11

11. PROBLEMS AFTER ORBIT IS ESTABLISHED

Temperature of Vehicle when on the Orbit. - The temperatures reached by the vehicle when it is on its orbit are calculated by considering the process of radiation of heat from the sun and earth to the body, and from the body to space.

Suppose, to begin with, that the simple illustrative example of the heating of the earth by the sun is considered. It is desired to compute the average temperature of the earth's surface. The rate of heat radiation by the sun will be

$$\frac{\partial Q}{\partial t} = \sigma T_s^4 \, \pi d_s^2 \tag{1}$$

where

$\sigma = 0.173 \times 10^{-8}$ BTU/(sq.ft.)(hr.)(deg.R.)^4

$T_s = 10,800^\circ$R., sun's surface temperature

$d_s = 864,000$ miles, sun's diameter

This energy travels into space on spherical surfaces, and therefore the fraction of this energy which is intercepted by the earth is equal to the percentage of the area of the sphere, of radius equal to the distance from the sun to the earth, which is blocked off by the earth. This must further be multiplied by the absorptivity of the earth. Therefore, the rate at which heat is absorbed into the earth is

$$\frac{\partial Q}{\partial t} = \sigma T_s^4 \, \pi d_s^2 \, \frac{\alpha A}{4 \pi L^2} \tag{2}$$

Chapter 11

in which α = absorptivity

A = projected area exposed to sun's rays

L = distance from earth to sun, 93×10^6 miles

The rate at which the earth radiates heat is

$$\frac{\partial Q}{\partial t} = \sigma \epsilon T_e^4 \pi d_e^2 \tag{3}$$

Here, ϵ = emissivity

T_e = average surface temperature of earth

d_e = diameter of earth, 7920 miles

When heat is radiated out as fast as it is absorbed, equilibrium conditions exist. This is obtained by setting (2) equal to (3).

$$\sigma T_s^4 \pi d_s^2 \frac{\alpha A}{4 \pi L^2} = \sigma \epsilon T_e^4 \pi d_e^2$$

From this is found, after setting $A = \pi d_e^2/4$, and $\epsilon = \alpha$ by Kirchoff's law,

$$T_e^4 = \frac{T_s^4 d_s^2}{16 L^2} \tag{4}$$

$$T_e = \frac{T_s}{2} \sqrt{\frac{d_s}{L}} \tag{5}$$

Using the values noted above, we find $T_e = 520°R = 60°F$, which is in reasonable agreement with our everyday experience.

Now let us apply a similar analysis to a satellite vehicle. The vehicle will alternately be in front and behind the earth. Supposing that it is in front of the earth for a sufficiently long time to reach equilibrium, the governing relation is

$$\sigma T_s^4 \pi d_s^2 \frac{\alpha A_{vs}}{4 \pi L^2} + \sigma T_e^4 \pi d_e^2 \frac{\alpha A_{ve}}{\pi d_e^2} = \sigma \epsilon T_v^4 S_v \tag{6}$$

Chapter 11

where the symbols are as noted previously except for

A_{ve} = projected area of vehicle as seen from earth

A_{vs} = effective projected area of vehicle as seen from sun

T_{vf} = temperature of vehicle, in front of earth

S_v = surface area of vehicle

Again assume $\alpha = \epsilon$, and simplify (6). T_e can be eliminated by the use of (4). This leads to

$$T_{vf}^4 = \frac{T_s^4 d_s^2}{16 L^2} \left[4 \frac{A_{vs}}{A_v} + \frac{A_{ve}}{S_v} \right] \tag{7}$$

Now it is necessary to determine the ratio of projected to surface areas. The vehicle is conical in shape, with an altitude of 16 2/3' and a base diameter of 3 1/3'. If the vehicle axis remains tangent to its orbit, then

$$\frac{A_{ve}}{S} = \frac{.5 \times 3.33 \times 16.67}{.5\pi \times 16.67 \times 3.33 + .25\pi \, 3.33^2} = .292$$

The projection of the vehicle as seen from the sun is a circle at "dawn", gradually changing to a triangle at "high noon" and then going back to a circle at "dusk". The mean fourth root of the projected area raised to the fourth power is

$$\frac{A_{vs}}{S_v} \doteq \frac{19.3 \times .292}{.5 \times 3.33 \times 16.67} = .203$$

Hence

$$T_{vf}^4 = 1.10 \frac{T_s^4 d_s^2}{16 L^2} = 1.10 \, T_e^4 \tag{8}$$

or

$$T_{vf} = 1.02 \, T_e = 530^\circ R = 70^\circ F.$$

Chapter 11

If the vehicle were behind the earth for a time great enough to produce equilibrium temperatures, then the equation governing the situation is like (6), but without the first term.

$$\sigma T_e^4 \pi d_e^2 \frac{\alpha A_{ve}}{\pi d_e^2} = \sigma \epsilon T_{vB}^4 S_v \tag{9}$$

This is readily reduced to

$$T_{vB}^4 = .292 \, T_e^4 \tag{10}$$

It follows directly that

$$T_{vB} = .732 \, T_e = 381°R = -79°F.$$

Thus we have shown that the temperature of the vehicle must lie between the limits 70°F. and -79°F. To find just where, between these limits, the temperatures lie, it is necessary to set up the differential equation relating temperatures with time. The net rate of heat transfer for the body in front of the earth with respect to the sun is

$$\frac{\partial Q}{\partial t}\bigg]_F = \sigma T_s^4 \pi d_s^2 \frac{\alpha A_{vs}}{4\pi L^2} + \sigma T_e^4 \pi d_e^2 \frac{\alpha A_{ve}}{\pi d_e^2} - \sigma \epsilon T_{v_F}^4 S_v$$

using (4), putting $\alpha = \epsilon$, and simplifying, we find

$$\frac{\partial Q}{\partial t}\bigg]_F = \sigma \epsilon S_v \left[1.10 \frac{T_s^4 d_s^2}{16 L^2} - T_{v_F}^4 \right] \tag{11}$$

For the values used previously,

$$\frac{\partial Q}{\partial t}\bigg]_F = \sigma \epsilon S_v \left[530^4 - T_{v_F}^4 \right] \tag{12}$$

The specific heat is defined to be

$$c = \frac{\Delta Q}{W \Delta T} \tag{13}$$

in which W is the weight of the vehicle. From this, it is seen that

$$\frac{\Delta Q}{\Delta T} = c W \frac{\Delta T}{\Delta t} \tag{14}$$

Chapter 11

(12) can then be reduced to

$$\frac{dT_{VF}}{dt} = \frac{\sigma \epsilon S_v}{cW}\left[530^4 - T_{V_F}^4\right] \tag{15}$$

There are several calculations which can be immediately made using (15). First, for equilibrium, $dT_{VF}/dt = 0$, and $T_{VF} = 530$. This checks the calculation made previously. Second, the greatest rate of heating occurs when T_V is least, and the least value of T_V is 381°R. Using that value, $\epsilon = 1.0$, $C = .12$ BTU/lb.$^\circ$F, and $W = 300$ lb*, we find $\partial T/\partial t = 265^\circ$F/Hr.

An expression similar to (15) can be set up for the time when the vehicle is behind the earth. For that case

$$\frac{dT_{VB}}{dt} = -\frac{\sigma \epsilon S_v}{cW}\left[T_{V_B}^4 - 381^4\right] \tag{16}$$

Here the maximum rate of increase in T_V is also 265°F./hr. Since the half-period is about .75 hr., it can be seen that the temperature changes will tend to be larger.

The mean temperature, T_{Vm}, can now be obtained by assuming that the rate of change of temperature is constant, and then

$$T_{V_m}^4 - 381^4 = 530^4 - T_{V_m}^4$$

From this we find that

$$T_{V_m} = 471^\circ R = 11^\circ F$$

An estimate of the limits to the temperature variation on the orbit can now be found by calculation of the rate of temperature change at this mean temperature.

---

* Structural weight only. Payload assumed to be insulated from structure.

Chapter 11

$$\frac{dT_{v_m}}{dt} = - \frac{\sigma \epsilon S_v}{cW} \left[471^4 - 381^4 \right]$$

and using the previously noted numerical values, we find the rate of change of T_V, at the mean T_V, is $132^{\circ}F$. per hour, or about $100^{\circ}F$ in the time required to make a half revolution.

To obtain the graph of temperature as a function of time, it is necessary to return now to (18) and (16). Inserting the values for the known constants, we have

$$\frac{dT_{v_F}}{dt} = (.821 \times 10^{-2})^4 \left(530^4 - T_{v_F}^4 \right). \tag{17}$$

and

$$\frac{dT_{v_B}}{dt} = - (.821 \times 10^{-2}) \left(T_{v_B}^4 - 381^4 \right). \tag{18}$$

For (17), we have

$$\int_{t_0}^{t_1} dt = \int_{T_{v_{F_0}}}^{T_{v_{F_1}}} \frac{dT_{v_F}}{(.821 \times 10^{-2})^4 (530^4 - T_{v_F}^4)}$$

Now

$$\frac{1}{a^4 - b^4} = \frac{1}{2a^2} \left[\frac{1}{a^2 - b^2} + \frac{1}{a^2 + b^2} \right]$$

$$= \frac{1}{2a^2} \left[\frac{1}{a^2 + b^2} + \frac{1}{2a} \left(\frac{1}{a-b} + \frac{1}{a+b} \right) \right]$$

$$= \frac{1}{4a^3} \left[\frac{2a}{a^2 + b^2} + \frac{1}{a-b} + \frac{1}{a+b} \right]$$

Hence

Chapter 11

$$t_1 - t_0 = \frac{1}{4(.321 \times 10^{-2})^4 530^3}\left[2\tan^{-1}\frac{T_{VF}}{530} + \ln\frac{530 + T_{VF}}{530 - T_{VF}}\right]_{T_{VF_0}}^{T_{VF_1}}$$

$$= .860\left(\tan^{-1}\frac{T_{VF_1}}{530} - \tan^{-1}\frac{T_{VF_0}}{530}\right)$$

$$+ .430\left(\ln\frac{530 + T_{VF_1}}{530 - T_{VF_1}} - \ln\frac{530 + T_{VF_0}}{530 - T_{VF_1}}\right) \tag{19}$$

when the vehicle is in the sun.

Equation (18) leads to

$$\int_{t_1}^{t_2} dt = \frac{-1}{(.821 \times 10^{-2})^4}\int_{T_{VB_1}}^{T_{VB_2}}\frac{dT_{VB}}{T_{VB}^4 - 381^4}$$

and

$$\frac{1}{x^4 - a^4} = \frac{1}{2a^2}\left(\frac{1}{x^2 - a^2} - \frac{1}{x^2 + a^2}\right) = \frac{1}{4a^3}\left(\frac{1}{x-a} - \frac{1}{x+a} - \frac{2a}{x^2 + a^2}\right)$$

so

$$t_2 - t_1 = -1.00\left(\ln\frac{T_{VB_2} - 381}{T_{VB_2} + 381} - \ln\frac{T_{VB_1} - 381}{T_{VB_1} + 381}\right)$$

$$+ 2.00\left(\tan^{-1}\frac{T_{VB_2}}{381} - \tan^{-1}\frac{T_{VB_1}}{381}\right)$$

when the vehicle is in the earth's shadow.

Equations (19) and (20) were used to obtain fig. 6 , which shows the time variations in temperature.

The calculations made up to this point have considered only radiation phenomenon. It is necessary also to investigate the heating due to friction. As in the case of the drag calculations, various regimes of heat transfer arise, depending primarily on the speed and altitude. The orbit will be of necessity chosen to be at an altitude so great that kinetic theory methods may be used here for computing heat transfer. We will begin with simple assumptions which will tend to make the heat transfer larger than it actually is.

DRM 25 BS
REV. 7.42

Analysis *TEMPERATURES*

Prepared by *HAROLD LUSKIN*

Date *5-6-46*

DOUGLAS AIRCRAFT COMPANY, INC.

SANTA MONICA   Plant

Page 177

Model

Report No. 11327

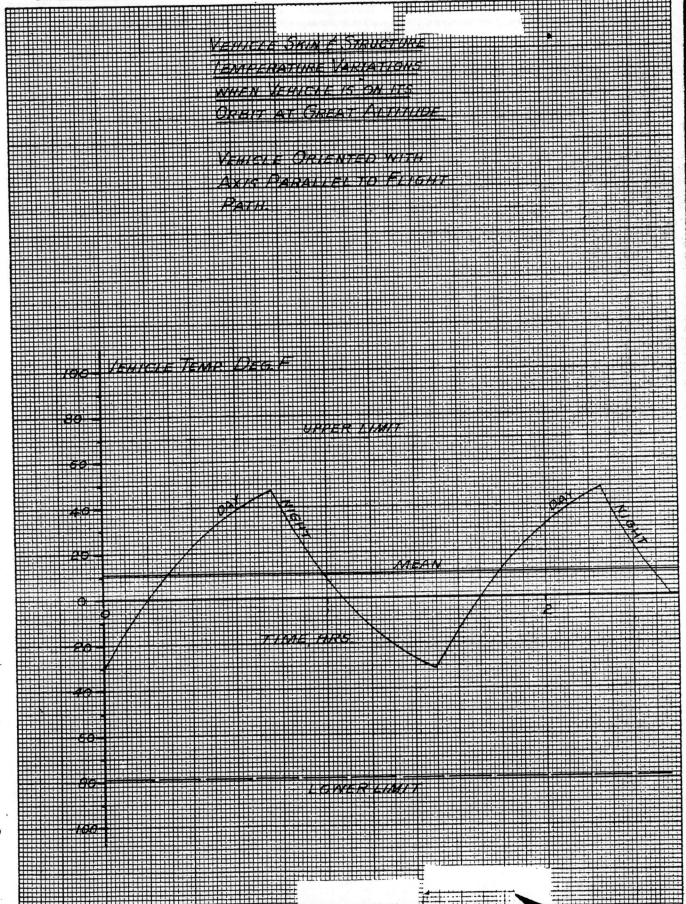

VEHICLE SKIN & STRUCTURE
TEMPERATURE VARIATIONS
WHEN VEHICLE IS ON ITS
ORBIT AT GREAT ALTITUDE

VEHICLE ORIENTED WITH
AXIS PARALLEL TO FLIGHT
PATH

Chapter 11

Consider a surface of the vehicle of unit area. Then if N is the number of molecules in unit volume, V is the vehicle speed, and α the angle between the surface and the direction of motion, the number of molecules colliding with unit area in unit time is $NV\alpha$. Each molecule has an energy $(1/2)\mu V^2$, where μ is the mass of one molecule, and where we disregard the random velocity because it is small compared to V. The energy of all the molecules which strikes unit area in unit time is evidently

$$E_{in} = \frac{1}{2}\mu N V^3 \alpha = \frac{1}{2}\rho V^3 \alpha \qquad (21)$$

The same number of molecules will leave unit area in unit time, but their momenta will be different. We will make the assumption that the air molecules enter the skin surface, come to thermal equilibrium with the skin molecules, and then are discharged in random fashion. The energy leaving the skin unit area in unit time is just

$$E_{out} = \frac{1}{2}\rho V c_W^2 \alpha \qquad (22)$$

where c_W is the molecular mean square speed corresponding to the skin temperature. The energy left in the wall is

$$\Delta E = \frac{1}{2}\rho V^3 \alpha \left[1 - \frac{c_W^2}{V^2} \right] \qquad (23)$$

Now by assuming equipartition of energy at the skin surface, we have

$$\frac{1}{2}M c_W^2 = \frac{3}{2}RT_W \qquad (24)$$

where M is the molecular weight of air, R is the universal gas constant and T_W is the temperature of the wall. Eliminating c_W between these

Chapter 11

equations leads to

$$\Delta E = \tfrac{1}{2}\rho V^3 \alpha \left[1 - 3 \frac{R}{M} \frac{T_W}{V^2} \right] \tag{25}$$

The most pessimistic case is that in which $3\frac{R}{M}\frac{T_W}{V^2}$ is ignored compared to unity. If, after making that assumption, the rate of heat transfer is found to be small, then since taking the additional term into account makes it yet smaller, the aerodynamic heating can be ignored. This assumption has the physical meaning that all of the energy of the oncoming molecules is left in the skin. Converting to BTU, we have the heat transfer into unit area in unit time,

$$\frac{\Delta Q}{\Delta t} = \tfrac{1}{2}\rho V^3 \alpha \frac{1}{J} \tag{26}$$

The following table gives values of the heat transfer at several altitudes and at orbital speeds. α is 0.1 radians. J is 778 ft. lbs. per BTU

ALTITUDE	DENSITY	SPEED	HEAT TRANSFER RATE
Miles	Slugs/cu.ft.	F.P.S.	BTU per sq.ft.per sec.
0	2.4×10^{-3}	25,900	3×10^{6} *
100	1.07×10^{-10}	25,600	1×10^{-1}
200	1.7×10^{-13}	25,300	3×10^{-4}
300	1.3×10^{-16}	25,000	1×10^{-7}
400	6.1×10^{-19}	24,700	6×10^{-10}

It is obvious that tremendous changes in heat transfer rate take place as altitude is changed. In order to get a scale with which to determine the importance of the aerodynamic heating, we calculate the radiation output of the same unit skin area that was considered above. This is

$$\frac{\Delta Q}{\Delta t} = \sigma \epsilon T^4$$

* Strictly speaking, (26) is not valid at sea level.

Chapter 11

which has a magnitude of 2×10^2 BTU per sq. ft. per second. It can be seen that the radiating power is extremely large compared to the rate of heat input, so long as the altitude is not below 200 miles. It seems justified, then, to neglect the aerodynamic heating for the cases of the cruising altitudes.

At the lower altitudes, where aerodynamic heating is important, more detailed expressions for aerodynamic heating are available, and these are discussed in other portions of this report.

It was assumed in these calculations, that day and night for the vehicle occupied equal times. That statement is approximately true only if the altitude is not great. The variation of length of daylight and night is shown in fig. 7.

The temperatures computed here are, in general, low compared to what we ordinarily think of as being "normal". There are several methods of controlling the temperature level, and raising it, if desired. First, if the missile is steered around its orbit so as to present its maximum projected area to the sun at all times when it is in the sunlight, a gain of 30°F. in mean temperature can be had. Second, reducing the emissivity will proportionately reduce the magnitude of temperature fluctuations, although the mean temperature will be unchanged. Third, surfaces of different values of emissivity (lower, away from the sun) will rise the mean temperature. Fourth, changing the plane of flight to one including the earth's axis and at the same time perpendicular to the sun at all times and raise the mean temperature to 70°F.

FORM 25 BS
(REV. 7-42)

Analysis __DAY & NIGHT__

Prepared by __LUSKIN & DANSKIN__

Date __5-7-46__

DOUGLAS AIRCRAFT COMPANY, INC.

SANTA MONICA   Plant

Page __181__

Model __#1033__

Report No. __11827__

HOURS OF SUNLIGHT AND DARKNESS FOR VEHICLE REVOLVING IN EARTH'S EQUATORIAL PLANE AT ORBITAL SPEEDS

HOURS PER REVOLUTION

SUNLIGHT

DARKNESS

ALTITUDE ABOVE SEA LEVEL, MILES

Chapter 11

11. PROBLEMS AFTER ORBIT IS ESTABLISHED - (Cont'd)

Aerodynamic Control while on Orbit. - The use of aerodynamic control is dependent on the ability to develop stagnation pressures (or indicated airspeed) of reasonably large order of magnitude. The stagnation pressure in turn, depends on the atmospheric density and the vehicle speed. In the present case, the orbital speeds are high, but at the desirable cruising altitudes the densities are extremely low. This follows directly from the fact that the densities must be kept low so that the drop per revolution is also low. The following table gives values of the indicated speed, V_i, for several altitudes, where

$$V_i = \sqrt{\rho/\rho_0} \; V$$

and where the stagnation pressure q, is

$$q = (1/2)\rho V^2 = (1/2)\rho_0 V_i^2$$

Here ρ_0 is standard sea level density.

ALTITUDE	SPEED	DENSITY	INDICATED SPEED
Miles	f.p.s.	Slugs/cu.ft.	f.p.s.
0	25,900	2.4×10^{-3}	25,900
100	25,600	1.07×10^{-10}	5
200	25,300	1.7×10^{-13}	0.2
300	25,000	1.3×10^{-16}	0.006
400	24,700	6.1×10^{-19}	0.0004

Now the force per unit area which can be developed by aerodynamic means is in the order of

$$\frac{F}{S} \approx .001 \, V_i^2$$

Chapter 11

At 100 miles altitude, F/S is about .025 lb., and for an area of 50 sq. ft. a force of 1.25 lb. would be developed. Thus, since the weight of the vehicle is about 1000 times this force, we see that it is hopeless to obtain accelerations or balancing of weight moments through aerodynamic means. The only use of such controls at high altitude would be to balance other aerodynamic forces or moments which, of course, would be of equally small magnitude.

Another factor of importance is that the lift drag ratios at high altitudes are in the order of one. To avoid losing altitude while maneuvering, it would be necessary to counteract the drag by thrust - the thrust being of same magnitude as the lift. It is obviously more economical to use the thrust force for lifting to begin with, and to dispense with the aerodynamic control.

Chapter 11

11. PROBLEMS AFTER ORBIT IS ESTABLISHED

Attitude Control by Recoil - After the rocket fuel is exhausted the jet rudders become inoperative and other means must be adopted to control the attitude of the missile on its orbit. In the extremely rarefied air of the ionosphere it may seem unnecessary to head the missile in the direction of its path because drag is negligible. However, if a definite orientation of the missile is desired, be it for purposes of orienting missile-borne instruments, of regulating temperature aboard or of guiding the missile on a descent into the lower atmosphere towards a slow-down and eventual landing, then means to turn the missile deliberately must be provided. Two such means have been proposed - viz, (1) missile-borne flywheels which when impelled will impart an equal and opposite angular momentum to the missile and (2) small torque rockets.

The feasibility of accommodating adequate flywheels in the satellite missile can be gleaned from the following study. Either in the instrumentation head or in the ring space around the thrust nozzle there appears to be room to accommodate three flywheels; one each for pitch, yaw and roll. The projectile weighing about 1000 lbs. empty is estimated to have a radius of gyration of 3 ft. An 8" diameter flywheel weighing about 2 lbs. would have to be rotated 50,000 times as much and as fast as the missile is to be turned in the opposite direction by reaction, provided no extraneous moments interfere. Spinning this flywheel at 2500 RPM would turn the projectile end for end in ten minutes. The power necessary to attain this speed in say 20 seconds, and to keep it running would be less than 1/50 HP. The braking could be done by friction. Less power would be required if more time is allowed or if the flywheel

Chapter   11

is made larger.

Actually there should be no pressing need for high rates of tilting the projectile. Much smaller rates should suffice for casual corrections. The average rate of rotation necessary to maintain the heading aligned with the path is one revolution in 95 minutes, about 1/5 of what the control device envisaged above could cope with. However, there is no reason why the orbital angular momentum should not be already imparted by jet rudder action during the last phase of the powered ascent before entry into the orbit.

As to roll control, the lesser moment of inertia of the vehicle in roll probably admits of similar attitude control with a smaller flywheel device than pitch and yaw control.

It would thus appear that a flywheel type of machine to influence the attitude of the satellite projectile in the absence of extraneous moments poses no serious problem from the viewpoints of bulk, weight, and power involved nor is the apparatus complicated or delicate.

Problems however, remain to be solved, to be sure. The precision with which orientation of the missile will have to be predicted or commanded remains to be determined and the methods by which the attitude of the projectile with respect to its path over the earth can be telemetered deserve serious study. Even though the air density at the orbit altitude is very small, aerodynamic disturbances may yet be commensurable with the control moments considered. If such aerodynamic moments are present for a large part of the time then their continued counteraction may entail the accumulation of large flywheel momentum.

Chapter 11

It would be possible to allow much larger maximum flywheel speeds than the 2500 RPM previously mentioned. Limitations are drawn by technical considerations of permissible flimsiness, because the larger a flywheel of a chosen weight, w, is made the longer it can absorb a given torque before rupturing its rim. The time integral of the reaction moment that can be put into the flywheel is $\int M\, dt = wL/\omega = wr\sqrt{L/g}$ where w and r are weight and radius of the rim while L is the breaking length of the material and ω the angular velocity eventually acquired by the flywheel. Assuming that the largest diameter flywheel of which a pair could perhaps be accommodated in the satellite missile is $2r = 32"$ and that the rim is made of steel having a breaking length of 67,000', then the product of reaction moment in ft.-lb. by the time in seconds that can be derived from each pound of rim weight is 60 ft.-sec. before the wheel would fly apart at 9600 RPM. Some factor of safety would have to be provided. For instance, a well-proportioned flywheel of 2 ft. diameter weighing 20 lbs. which would be allowed to attain 10,000 RPM would cope with a systematic average aerodynamic moment of 1/100 ft.-lb. for 10 orbit periods in its plane. This amounts to 2 lbs. of wheel weight investment per orbit.

While rotors are spinning there will be gyroscopic cross influences which may require mutual corrective devices but they are not accumulative over more than half a satellite period.

The second proposal of attitude control by reaction envisages small gas exhaust recoil guns. Six or eight such devices arranged to emit jets tangentially to the skirt of the projectile can provide control

Chapter 11

of rotation in pitch, yaw and roll.

If the gun were mounted to fire radially at a leverage ℓ from the center of gravity of the projectile of weight W and radius of gyration i, it would have to develop a recoil force of $P = \pi n W i^2 / 30 \, g \ell t$ in order to accumulate a pitching rate of n RPM in t seconds. Assuming $\ell = 4$ ft., $W i^2 / g = 300$ slugs ft.^{2} and $n = 1/20$ as before, $Pt = \pi/8$ lb.-sec. In other words, a recoil of about 6 ounces acting for one second or 1/10 ounce acting for one minute would suffice to achieve the specified effect. The amount of gas to be discharged at supersonic velocities under some pressure would be of the order of 1/10 oz. each time such control is given. Such gas could be branched off from an existing nitrogen pressure system or from an alcohol oxygen burner or even an oxygen vaporizer.

Again the question of how much aerodynamic pitching moment the control may have to cope with remains to be investigated. If again 1/100 ft.-lb. average torque were to be generated continuously, then only 1/4 lbs. of gas per circuit of orbit would have to be expended. On the first glance this looks about 8 times as favorable as the flywheels, but considering that momentum is no longer conserved for moment reversals the two schemes appear essentially on a par as far as weight investment is concerned. Other means may be considered for creating erecting moments or for keeping the disturbing moments small. One of the most drastic ones would be to house the real "payload" instruments in a spherical shell and expel it from the satellite vehicle after it is established in the orbit.

11. **PROBLEMS AFTER ORBIT IS ESTABLISHED.** - (Cont'd)

<u>Loss in Height Due to Resistance while on Orbit.</u> - Preliminary

calculations on the loss of altitude as time goes on showed that at altitud-

es of about 100 miles only small rates of drop would be encountered.

Such calculations were based on the following formula, obtained by small

perterbations from the basic equations of motion.

$$\frac{\Delta R}{R} = 4\pi \frac{D}{W}$$

ΔR is the loss in altitude per revolution, R is the orbital radius, D

is the drag, and W the weight. For a hasty estimate of drag one is tempted

to say $C_D \sim \dfrac{1}{\sqrt{M^2-1}} \sim \dfrac{1}{M}$, and then $D = \dfrac{1}{M} \rho \dfrac{V^2}{2} A$

Here M is the Mach number, ρ the air density, V the speed of the vehicle,

and A its frontal area. To arrive at a magnitude for ρ, we assume an

isothermal atmosphere. From the condition for equilibrium of the atmos-

phere, we have

$$dp = -\rho \, g \, dh$$

where p is the pressure, and h the altitude. From the gas law, we have

$$dp = d\rho \, g \, R_g T$$

where R_g is the gas constant.

Combining these two equations to eliminate the pressure yields

$$\frac{d\rho}{\rho} = - \frac{dh}{R_g T}$$

Integration then leads to

$$\log \rho = - \frac{h}{R_g T} + \text{const.}$$

The constant can be evaluated by saying $\rho = \rho_0$ when h = o.

Chapter 11

Then $\qquad \log \rho - \log \rho_o = - \dfrac{h}{R_g T}$

This can be written

$$\rho = \rho_o e^{-\frac{h}{R_g T}}$$

Suppose, now, the following numerical values are assigned to the quantities in this relation.

$R = 4,000 \times 5,280$ ft.

$W = 1150$ lbs.

$A = 8.7$ sq. ft.

$V = 26,000$ ft. per second

$M = V/\sqrt{\gamma g R_g T} = 26,000/\sqrt{1.4 \times 53.3 \times 500 \times 32} = 24$

$\rho_o = .002378$ slugs/cu. ft.

$h = 100$ miles $= 528000$ ft.

$T = 500^\circ$ R.

The density can easily be computed to be $.48 \times 10^{-11}$ slugs per cu. ft., and the drag is .00059 lb. The drop per revolution is then found to be 135 ft. Doubling the altitude, making it 200 miles, would have the effect of squaring the value of ρ/ρ_o, and finally changing ΔR to about 10^{-9} feet per revolution.

These results would seem to indicate that the loss in altitude was sufficiently small at altitudes well above 100 miles. In the first place, due to the extremely low Reynold's numbers encountered at high altitudes, the drag must be evaluated by unconventional methods. Secondly, the assumption of an isothermal atmosphere can carry with it large errors.

Chapter 11

For these reasons, a more comprehensive study was undertaken.

A re-evaluation of the drag coefficients, based on the principles of the kinetic theory of gases was made and is presented in Appendix B. The revised drag coefficients were found to be considerably higher. This increase is seen to be reasonable when the known variations with Reynolds number are taken into account. In addition, a search of the literature was made in order to arrive at better estimates of the density of the atmosphere. The results of this study are presented in Appendix A. The following table sums up the corrections to the above method which can be made.

Altitude, miles	0	100	200	300	400
Temperature, deg. Rankine	519	906	1670	774	671
μ, viscosity, lb.sec./sq. ft.	3.7×10^{7}	5.5×10^{7}	8.3×10^{7}	4.9×10^{7}	4.4×10^{7}
ρ, density, lb. sec.^{2}/ft.^{4}	2.4×10^{3}	1.07×10^{10}	1.7×10^{13}	1.3×10^{16}	6.1×10^{19}
V, orbital speed, fps	25,900	25,600	25,300	25,000	24,700
V_i, $V\sqrt{\rho/\rho_o}$, indicated speed, fps	25,900	5	0.2	0.006	.0004
N_R, Reynolds no.	2×10^{9}	6×10^{1}	6×10^{2}	8×10^{5}	4×10^{7}
M, Mach no.	23.2	17.5	12.7	15.7	14.5
C_D, drag coefficient	.2	1.0	2.0	2.0	2.0
Drag, lbs.	3×10^{6}	6×10^{1}	2×10^{3}	2×10^{6}	1×10^{8}
ΔR, drop per rev., ft.	7×10^{11}	1×10^{5}	5×10^{2}	5×10^{1}	2×10^{3}
Approx. time to drop to earth	0	1 hr.	3 weeks	23 yrs.	10 centuries
Approx. range before dropping to earth, mi.	0	1.75×10^{4}	8.7×10^{6}	3.43×10^{9}	1.47×10^{11}
Cycles before dropping to earth	0	6.9	3330	1.28×10^{6}	5.37×10^{7}
Time to drop to earth if ρ is in error by a factor of 1000	0	0	.5 hr	8.5 days	1 yr.

Chapter 11

The viscosity data were obtained, as functions of temperature, from McAdams, Heat Transmission, Second Edition, p. 411, and Durand, Aerodynamic Theory, Vol. 6, p. 227. Using these data, it is found that ΔR is grossly different than was calculated by the first, erroneous, set of assumptions.

The primary cause of the large difference between the two calculations is the difference between the isothermal and the actual atmosphere. The large temperatures existing at great altitudes causes the density to drop off much more slowly than would have been predicted by using lower temperatures. The following table, for example, shows the actual and isothermal (500°R) values.

ALTITUDE	ρ (T=500°R)	ρ(ACTUAL)
100 mi.	4.8×10^{-12}	1.07×10^{-10}
200 mi.	2.3×10^{-25}	1.7×10^{-13}

These differences are the direct result of the type of temperature variation assumed in each case. It will be noticed that the revised values shown in the table indicate that a vehicle starting in an orbit at an altitude of 200 miles would remain aloft about 3 weeks. However, when the accuracies of the assumption underlying these calculations are examined critically, particularly the values of density in the upper atmosphere, it is found that the duration figures may yet be in error by a factor of 1000. For this reason an additional line is included in the table showing the lower limit believed possible for the duration. Confronted by our present state of ignorance, one can only conclude that the minimum initial altitude for a satellite should be 200 miles while the recommended altitude would be between 300 and 400 miles.

Chapter 12

12. THE PROBLEM OF DESCENT AND LANDING

An important ultimate goal for any vehicle must be that of carrying human beings with safety. One obstacle which seems to stand in the way in the present case is the great energy stored in the vehicle, a part of which serves to heat the vehicle on descending into the lower atmosphere. The study which follows is an attempt to show the feasibility of lowering the craft without destroying it by fire, so that a safe landing can be made on the surface of the earth.

Landing introduces primarily a problem of dissipation of the tremendous energy stored in the vehicle by virtue of its speed. If all of this energy, for example, were to be converted to sensible heat of the vehicle, the temperature would be increased by

$$\Delta T = \frac{\frac{1}{2} \frac{W}{g} V^2}{c W J} \quad \text{-------------------- (1)}$$

where:

W = vehicle weight, lbs.

g = acceleration of gravity, ft./sec./sec.

V = vehicle speed, ft./sec.

c = specific heat of vehicle, BTU/lb./$^{\circ}$F.

J = 778 ft. lbs./BTU

choosing $g = 32.2$, $V = 26,000$, $c = .12$, leads to $\Delta T = 112,000$ $^{\circ}$F. The time required to radiate all this energy into space can be estimated as follows:

$$\Delta t = \frac{\frac{1}{2} \frac{W}{g} V^2}{\sigma \epsilon T^4 A J} \quad \text{---------------------- (2)}$$

where:

σ = Stefan-Boltzmann constant, BTU/sq. ft./hr./($^{\circ}$F.)^4

ϵ = emissivity

T = vehicle temperature, $^{\circ}$R.

A = vehicle surface area, sq. ft.

choosing $\sigma = .173 \times 10^{-8}$, $\epsilon = 1.0$, and $A = 95$, $W = 1150$, leads to $\Delta t \, T^4 = .940 \times 10^{14}$. If the vehicle is allowed to radiate at as

Chapter 12

high a temperature as 1000°R., then 94 hours would be required for the necessary removal of heat energy. The burning of meteors as they enter the earth's atmosphere is a striking example of the type of phenomenon which might be expected.

Actually the dissipation times will be only a small fraction of that computed above because a large portion of this energy is left behind the vehicle in the form of heat in the wake. The remainder is radiated from the surface, is stored in the vehicle as heat, or is retained as kinetic energy to be dissipated during the landing run on the ground. We deal here only with that energy which is fed into the vehicle as heat, and which must be radiated back into space.

The solution to the problem must come from a method of controlling the trajectory during the landing glide so that the heat input can be dissipated at a temperature sufficiently low to prevent damage to the vehicle. Such control of the glide path must be accomplished by aerodynamical means, implying that lifting surfaces must be provided to prevent the vehicle from entering the denser region of the atmosphere too rapidly. The trajectory will now be investigated by studying the heat flow balance.

The heat input from the boundary layer of a cone is

$$\frac{\partial Q}{\partial t} \frac{1}{S} = H(T_{BL} - T_s) \tag{3}$$

Chapter 12

where:

Q = heat energy, BTU

t = time, sec.

S = surface area, sq.ft.

H = heat transfer coefficient, BTU/$^{\circ}$F./sq.ft./sec.

T_{BL} = boundary layer temperature

T_s = vehicle surface temperature

The heat transfer coefficient has been shown to be

$$H = .0224 \frac{\lambda}{\ell} \left(\frac{V\ell}{\nu}\right)^{0.8} \beta^{1/3} \tag{4}$$

in which

λ = heat conductivity of air at the temperature T_{BL}, BTU/sec/ft/$^{\circ}$F.

ℓ = cone length, ft.

V = velocity of vehicle, ft/sec.

$\nu = \mu/\rho$

μ = viscosity of air at temperature T_{BL}, lb.sec/sq.ft.

ρ = density of air at temperature T_{BL}, slugs/cu.ft.

β = total cone angle, rad.

The work which follows will be much simplified by the arbitrary assumption that the exponent of the Reynolds number be changed from 0.8 to 1.0. (4) then becomes

$$H = .0224 \frac{\lambda}{c_p \mu} \, c_p \rho \, \beta^{1/3} V \tag{5}$$

where:

c_p = specific heat of air at constant pressure, BTU/lb/$^{\circ}$F.

Chapter 12

The group $\lambda/c_p\mu$ is the reciprocal of the Prandtl number, and is approximately constant. Using 0.7 for $c_p\mu/\lambda$, 0.24 for c_p, and .2 for ρ, we find

$$H = .0045\ \rho V \tag{6}$$

(6) can now be combined with (3), yielding

$$\frac{\partial Q}{\partial t} = .0045\ \rho VS(T_{BL} - T_s) \tag{7}$$

Now it is well-known that the stagnation temperature is related to the Mach number by the relation

$$T = T_h(1 + .2M^2) \tag{8}$$

Here we use T_h for the ambient air temperature. Experience based on German wind tunnel results has shown that the boundary layer temperature is somewhat less than the stagnation temperature, so we can write

$$T_{BL} = T_h(1 + .18M^2) \tag{9}$$

This relation, combined with (7) gives the final expression for the rate of heat input.

$$\frac{\partial Q}{\partial t} = .0045\ \rho VS\ (T_h + .18M^2 T_h - T_s) \tag{10}$$

The effect of the sun is not important here, as is shown by comparison to the flight of meteors, which become very hot only on passing through a gaseous atmosphere.

Chapter 12

For the heat output, we have the familiar radiation equation,

$$\frac{\partial Q}{\partial t} = \epsilon \, \sigma S T_s^{4} \qquad (11)$$

in which

  ϵ = emissivity

  σ = Stefan-Boltzmann constant

Now we shall define a special trajectory of descent, one in which a constant temperature of the vehicle is maintained. That assumption implies, as is shown by reference to equation (11), that a fixed rate of heat transfer is characteristic of this trajectory. This fixed rate of heat output must be matched by an equal heat input by proper choice of speed at each altitude in order that the skin temperature not change. The analytical expression of this trajectory is obtained by equating (10) and (11)

$$.0045 \, \rho V \left(T_h + .18M^2 T_h - T_s \right) = \epsilon \, \sigma T_s^{4} \qquad (12)$$

For any given temperature of the vehicle, (12) defines a unique speed for each altitude. Curves showing these glide paths for several vehicle temperatures are on figure 1.

There are several important results to be obtained from figure 1 . First of all, since the contours of fixed skin temperature are essentially horizontal, no serious heating due to friction will be encountered above the 70 mile level, at least for speeds in the general order of 25,000 feet per second. Second, at heights below

Analysis _LANDING_

Prepared by _A.S._

Date _5-6-46_

DOUGLAS AIRCRAFT COMPANY, INC.

S.M.

Plant

Page _197_

Model _# 1033_

Report No. _11827_

TRAJECTORY REQUIREMENTS TO MAINTAIN
CONSTANT TEMPERATURE ON VEHICLE DURING
DESCENT THROUGH THE EARTH'S ATMOSPHERE

ORBITAL VELOCITY

$T_S = 659° R$

$T_S = 759° R$

$T_S = 959° R$

$T_S = 1459° R$ (INCIPIENT RED HEAT)

ALTITUDE (MILES)

- 70
- 60
- 50
- 40
- 30
- 20
- 10
- 0

ALTITUDE (FEET)

- 300,000
- 200,000
- 100,000
- 0

VELOCITY (F.P.S.)

- 5000
- 10,000
- 15,000
- 20,000
- 25,000
- 30,000

Chapter 12

70 miles, the temperature rises to very high values, unless the speed is reduced along with the reduction in altitude. To make a successful glide, therefore, it is necessary to be able to control the speed. We shall assume, then, that wings of small size will be used for speed control during descent, and for making landings. It is important to notice at this point that once sufficient lift is provided to control the speed, the effect of high drag is to reduce the time of descent. Arrangements with poor values of lift drag ratio are not, for that reason, out of order here.

The question arises as to just what size wing is required. The answer to this question comes from considering the vertical forces acting. If wings are used to slow down below orbital speeds, then a lift on the wings must be developed which is equal to the difference between weight and centrifugal force.

$$L = (1/2)\rho V^2 C_L S' = W - \frac{W}{g}\frac{V^2}{R} \tag{13}$$

where:

L = lift

S = wing area

R = radius of orbit $\doteq$ earth's radius

W = vehicle weight

ρ = density of atmosphere

V = vehicle speed

C_L = lift coefficient

g = acceleration of gravity $\doteq$ 32 ft/sec.^{2}

Chapter 12

The parameter $C_L S$ can readily be computed now for each of the trajectories of figure 1. The results of these calculations are given in figure 2. Maximum values of $C_L S$ as a function of skin temperature are given in figure 3. The vehicle weight was taken to be 800 lbs. for these calculations.

An additional consideration which must be taken into account, is the landing speed. Here $C_L S$ is a function of the desired landing speed according to

$$C_L S = \frac{2W}{\rho V^2}$$

For a landing at sea level, the landing speeds corresponding to several values of $C_L S$ have been noted in figure 3.

Inspection of figure 3 shows that wing areas compatible with reasonable skin temperatures and modest landing speeds are possible of achievement. For example, a landing speed of 100 mph, a maximum temperature of 300°F., (existing for only a brief section of the glide path) are consistent with a wing area of 30 sq.ft., (for $C_L = 1.0$) even if body lift is ignored. This corresponds to a 27 lb. wing loading.

It can be concluded as a result of this study that it appears possible to glide a space vehicle down to a landing on the earth's surface without destruction by fire or from crash landing. This gives rise to a hope of attempting space flights in man carrying craft.

FM 23 85
REV. 7-42

Analysis __LANDING__

Prepared by __A S__

Date __5-1-46__

DOUGLAS AIRCRAFT COMPANY, INC.

SANTA MONICA

Plant

Page __200__

Model __#1033__

Report No. __SM11827__

LIFTING SURFACE AREA REQUIRED

TO MAINTAIN CONSTANT TEMPERATURE
OF VEHICLE DURING DESCENT THROUGH
THE EARTH'S ATMOSPHERE

ϵ EMISSIVITY = 1.0
T_S, SURFACE SKIN TEMPERATURE = AS NOTED
C_L, LIFTING SURFACE LIFT COEFFICIENT
S, LIFTING SURFACE AREA

$T_S = 659°R$

$T_S = 759°R$

$T_S = 959°R$

$T_S = 1459°R$

ALTITUDE
(FEET)

300,000

200,000

100,000

0

$C_L S$ (SQUARE FEET)

0   10   20   30   40   50   60

MAXIMUM LIFTING SURFACE REQUIREMENTS TO
MAINTAIN CONSTANT TEMPERATURE OF VEHICLE
DURING DESCENT THROUGH EARTH'S ATMOSPHERE

LANDING SPEED AT SEA LEVEL NOTED ON CURVE

FIGURE E. CHAPTER 13

$C_L S_{MAX.}$ (SQ. FT.)

$T_{MAX.}$ °F

100 FPS

150 FPS

200 FPS

250 FPS

1000

500

0

0   10   20   30   40   50   60

Chapter 12

The ascent to altitude differs from the descent in one important way. The climb starts at zero velocity so that the region of high densities is passed through before great speeds are built up. In the descent on the other hand, the regions of high density and of high heat transfer coefficient are traversed at very high speed. This fact makes for small gains in temperature during the climb. For a skin thickness in the order of 0.1 inches, a temperature rise of less than one hundred degrees Fahrenheit has been computed.

Chapter 13

13. DESCRIPTION OF VEHICLE

A considerable amount of study has been devoted to the design of a practical satellite vehicle based upon the principles outlined in the preceding chapters. Two alternatives have been considered, one a four stage rocket fueled with alcohol, the other a two stage rocket fueled with hydrogen. Of these the former is lighter, smaller and by far the more conservative in terms of research requirements, safety and certainty. It will be described in some detail as to stages and technical components. The hydrogen vehicle is a much more speculative project, dependent on uncertain outcome of more ambitious and dangerous research projects. Its design aspects are therefore only briefly outlined in the last section of the present chapter for comparison. The three pictures shown at the beginning of the present chapter will give an idea of what the vehicles and their components are expected to look like. A rough weight breakdown of the two projects is presented in Chapter 7.

Shape and Dimensions. The shape will be that of a typical projectile, having a pointed nose and contoured sides which taper from a maximum width aft of the midsection to a minimum width at the base which is compatible with the motor exit diameter and space requirements for jet-vane controls. The length will be of the order of 60 to 70 feet and diameter about 12 to 14 feet, giving a fineness ratio between 4:1 and 6:1. Such a vehicle would have a density ratio, .029 pounds of loaded missile per cubic inch of volume, comparable to the German V-2 ratio of approximately .020. This indicates considerable progress

ORM 25-S-1
(REV. 8-43)

PREPARED BY: E. Baker   DOUGLAS AIRCRAFT COMPANY, INC.   PAGE: 204

DATE: May 2, 1946   SANTA MONICA   PLANT   MODEL: #1033

TITLE: PRELIMINARY DESIGN OF SATELLITE VEHICLE   REPORT NO. SM-11827

Chapter 13

towards better utilization of space.

Stages.   The vehicle will be divided into four stages, the primary or first stage being nicknamed "Grandma", the second stage "Mother", the third stage "Daughter", and the final satellite vehicle "Baby". Baby will carry the payload and intelligence for all stages, in addition to its own fuel, pumps, motor and guidance, and will comprise between 1/5 and 1/4 of the length of the total vehicle. Daughter and Mother will each carry only fuel, pumps, motors, and controls, being guided by Baby, and will be about the same length as Baby. Grandma will comprise almost half the length of the total vehicle, and will also contain fuel, pumps, motor, and controls, being guided by Baby.

Structure.   From the standpoint of size and applied loads, this vehicle is not out of proportion to present day large airplanes of 50,000 lb. gross and over, so it is believed entirely feasible to use airplane type of construction consisting of reinforced sheet metal.   The possibility of elevated temperatures existing on the surface will probably eliminate the use of aluminum alloys for skin covering and require the use of high strength stainless steel. Motor loads in any stage may be carried into the outer skin of that stage by a truncated cone diaphragm extending from the motor base to the outer skin. The thrust load of any stage can be transmitted into the next succeeding stage by pads on either the motor base or outer covering of the next stage.   The base of Grandma must be amply reinforced to withstand the dead weight of the entire assembly when mounted vertically ready for launching.

Although all flight loads are essentially directly along the axis of the vehicle, some reinforcing may be required to account for handling loads while each unit is in a horizontal position.

Tankage.   In the event that the oxidizer tank does not need to be insulated from the missile outer skin, the skin will serve as both tank shell and missile shell. If insulation becomes necessary, a double skin thickness will be required, thereby increasing the length or diameter, or both, of the entire vehicle. The front diaphragm of the forward tank in each stage must be reinforced to withstand internal pressures created by excess of vapor pressure over outside air pressure at higher altitudes. When liquid oxygen is used as an oxidizer, some form of insulation will probably be required between the two tanks. This will require a double diaphragm between the two tanks. All diaphragms can be of a near-ellipsoidal shape to more nearly approach the optimum weight vs. volume ratio under the existing hydrostatic head.

Motor.   The motors are designed for liquid oxygen and alcohol as the propellants with a mixture ratio of 1.5 (wt. of oxidizer/wt. of fuel.) The motor of the fourth stage is patterned after the light weight type developed by the Jet Propulsion Laboratory of C.I.T. The motors of 1st, 2nd, and 3rd stages are patterned after the German throatless motor which was intended for later use in the V-2 and Wasserfall. It consists primarily of the divergent section of the conventional motor with a short combustion section the same size as the throat added to the front. The injector is of completely new design which gave good combustion very

Chapter 13

close to the face of the injector. With but a very small penalty in gas velocity this new type of combustion chamber was adopted in the design of the motors for Stages 1, 2 and 3.

All the motors are regeneratively cooled with the Alcohol being used as the cooling medium. The Alcohol is brought to the exit section of the motor where it enters the cooling coils for circulation around the motor. The Alcohol is then directed to the injector head for injection into the combustion chamber.

The nozzle expansion ratio (area of exit/area of throat) is 5.0 for the first stage and 20 for all succeeding stages. The nozzle is shaped like a bell jar for rapid expansion near the throat and very little expansion near the exit. This will direct the jet flow straight to the rear and will permit the location of the jet rudder control inside the nozzle exit.

Fuel System. The fuel system is visioned as comprising a dual pump arrangement for delivering the fuel and oxidizer from the tanks to the motor through the necessary control and regulator valves. The fuel can be used as coolant for the motor by passing through coils arranged either helically or radially around the motor. Under this arrangement, both fuel and oxidizer pumps would be mounted on a common shaft and so designed that proper mixing ratios would be maintained at all operating speeds. The pump shaft would be driven by a steam turbine. Past practice has been to use the action of a catalyst on hydrogen peroxide to generate steam for the turbine. The hydrogen peroxide is delivered to

ORM 25-S-1 (REV. 8-43) PREPARED BY: B. Baker	DOUGLAS AIRCRAFT COMPANY, INC.	PAGE: 207	
DATE: May 2, 1946	SANTA MONICA PLANT	MODEL: #1033	
TITLE: PRELIMINARY DESIGN OF SATELLITE VEHICLE		REPORT NO. SM-11827	

Chapter 13

the steam generator by nitrogen gas under high pressure.

In some cases it might be necessary to use multi-stage pumps to prevent cavitation, but all pumps must be designed for extreme light weight by the use of hollow blades and other devices. They must also operate close to peak efficiency at design speed and at lower throttled speeds. The design of both turbine and pumps will be facilitated by the short duration of run.

Various on-off valves must be provided to start each turbine pump at the proper time, and some means of throttling to prevent excessive vehicle acceleration must be incorporated in these valves. A control valve operated by the motor coolant must be included in the oxidizer line to prevent the oxidizer from prematurely reaching the motor and causing an explosion.

Plumbing to the motor must be provided with expansion joints to prevent breakage of lines.

Controls. The control system of the missile comprises attitude and thrust controls for each stage and a common regulator system which governs the controls from a central brain station located in the final Baby unit. The attitude control is effected by a cruciform array of four vanes mounted on radial shafts near the rim of the motor nozzle. They are tilted on radial shafts by servo motors so as to deflect the hot rocket gas jet. By symmetrical deflection pitching or yawing re-action moments are developed. To create a rolling moment a cyclic differential deflection is superimposed. Thrust is controlled by regu-

Chapter 13

lating the fuel pumping speed and governed by accelerometers.

The master regulator in the "brain station" comprises an automatic
pilot system based on a tri-axial reference gyroscope system and a
tilting program governor which is designed to follow a pre-set ascent
trajectory calculated to enter the orbit smoothly. A radio altimeter
as well as radio guide beam or remote command receivers are installed
in the Baby missile. Their outputs are mixed with those of the automatic regulator so as to permit overriding the latter and applying
corrections for unforeseen disturbances.

The Baby vehicle is further equipped with small reaction motors
designed to exercise a moderate amount of control of orientation of
the projectile when in the orbit.

Telemetering and beacon equipment will supplement the control
system by maintaining transmission of tracking and intelligence information to the ground director station.

Accessories. Among accessories carried on the missile are: electric power plants to supply electric power to the instrumentation,
regulators, and servosystem; stage separation devices and their controls;
safety devices to interlock various functions; and appliances necessary
for fueling, readying and testing before take-off.

Payload. The nose cone of the Baby vehicle is reserved for the
payload and meant to house most of the useful instrumentation which is
to be carried on the journey to secure information gathered there. The
payload cone dimensions as now visualized are about 3 ft. diameter of
the bulged base, and 7 ft. length. It houses about 20 cu. ft.; 500 lbs.

Chapter 13

are alotted to its content as payload proper. Some of the payload items may optionally be accommodated in the annular space around the thrust motor of the Baby vehicle.

Operation. The operational procedure in launching a satellite missile is visualized essentially in the following manner: All parts of the vehicle after having been proof tested are brought to the launching site where a base with a blast apron and suitable scaffold have been installed. Here the complete vehicle is assembled upon a launching stand by progressively hoisting the lesser stages on top of the larger ones. All systems are checked as each stage is completed. Next all containers are filled from bottom to top and eventually topped off, and the hoist moved away to clear. In the meantime all ground observation stations are placed, manned and warned. Actual firing is triggered remotely from a protected control station according to pre-arranged schedule.

Hydrogen Alternative. The alternative project of a hydrogen fueled vehicle as shown in an exploded view in the last of the three pictures heading the present chapter, would differ from the alcohol fueled version in that it would be much larger, that it would be a two-stage affair and that elaborate precautions against the dangers of hydrogen leakage and evaporation would have to be taken. This is indicated in the picture by showing a double wall around the liquid hydrogen tank. Whether the oxygen would eventually be carried above the hydrogen as shown or rather below it would have to be decided after research will have clarified the conflict between advantages and dis--

advantages of either system, but the choice would hardly affect the general appearance of the assembled missile. The operational problems of liquid hydrogen fuel cannot be foreseen as well as those of the alcohol fuel.

Chapter 14

14. POSSIBILITIES OF A MAN CARRYING VEHICLE

Throughout the present design study of a satellite vehicle, it has been assumed that it would be used primarily as an uninhabited scientific laboratory. Later developments could alter its capabilities for use as an instrument of warfare.

However, it must be confessed that in the back of many of the minds of the men working on this study there lingered the hope that our impartial engineering analysis would bring forth a vehicle not unsuited to human transportation.

It was of course realized that 500 lbs. and 20 cubic feet were insufficient allotment for a man who was to spend many days in the vehicle. However, these values were sufficient to give assurance that livable accommodation could be provided on some future vehicle.

The first question to be considered in determining the possibility of building a man carrying vehicle is whether prohibitively high accelerations can be avoided during the ascent. The V-2 gave hope that this was possible. Our own studies have likewise shown that the optimal accelerations do not exceed about 6.5g. A man can withstand such acceleration for the periods of time involved (several minutes) if he is properly supported with his trunk lying normal to the direction of the acceleration. In Chapter 8, it will be remembered, the analysis showed that the performance could be improved a small amount by throttling each rocket motor during the latter portion of its burning period in order to reduce the structural loads. Under these conditions, the maximum accelerations could be profitably reduced to about 4 g. All these findings confirm

PREPARED BY: F. H. Clauser	DOUGLAS AIRCRAFT COMPANY, INC.	PAGE: 212
DATE: May 2, 1946	SANTA MONICA PLANT	MODEL: #1033
TITLE: PRELIMINARY DESIGN OF SATELLITE VEHICLE		REPORT NO. SM-11827

Chapter 14

that ascent offers no insurmountable obstacle to the construction of an inhabited satellite vehicle.

Next we consider the safety and welfare of the man after the vehicle has been established on the orbit. Popular fiction writers have devoted considerable thought and ingenuity to means of furnishing him with air, food and water. The most ingenious of these solutions is that of the balanced vivarium in which plants and man completely supply each others needs. Leaving these problems to the inventors, we ask ourselves the engineering questions of whether we can provide livable temperatures and a reasonable protection against meteors. In Chapter 11 we have seen that the answers are tentatively in the affirmative.

Lastly we consider the problem of safely returning the vehicle's inhabitant to the surface of the earth. In Chapter 12, we have seen that, with reasonable area wings, we can control the descent sufficiently to avoid dangerously high temperatures. These same wings are adequate to accomplish the final landing on the earth's surface.

The above thoughts are far from final answers on this problem. However, they do give a note of assurance that the hope of an inhabited satellite is not futile.

Chapter 15

15. ESTIMATION OF TIME AND COST OF PROJECT

Estimate of Time Required for Project - The progress of the preliminary design study presented in this report has indicated it is extremely important that at least six months additional research and preliminary design work be done on this project before any definite design and building program be established. It is possible by doing this that a three to five year program could be planned which would take into consideration the technological advances expected during this period so that obsolescence would not overtake the development.

The above statement and those which follow in the sections on Program and Costs are based on the requirements for the alcohol-oxygen rocket units, and not on the hydrogen-oxygen alternate. Estimates on the hydrogen-oxygen units are virtually impossible at this time due to the unknown factors in the use of liquid hydrogen.

Chapter 15

Program of Building and Testing. In view of the complexity of
the job and the increase of its magnitude with the transition to larger
launching stages, which for brevity's sake will be referred to as Baby,
Daughter, Mother and Grandmother, it will be desirable not to tackle
all stages at once but rather to progress from the smaller to the larger
in sequence in order to reap the full benefit of experience gained as
the job moves along. At present the extrapolation from prototype to
unprecedently large mother stages can, by the very nature of such a process, only be delicate and groping. The actual size of the real article
sensitively depends on relatively small changes in technological assumptions. This apparent uncertainty will disappear as successive tests of
the lesser stage units will supply the information upon which the design
of the larger ones can be solidly based. The smallest baby and daughter
stages however are of conventional dimensions well within existing practice and experience, so their design can well be immediately begun to-
gether in order to accelerate solution of the birth-aloft problem which
is the most drastic innovation over the existing practices. The manufacturing program will be overlapping as dictated by the testing program.
It is anticipated that more of the smaller stage units will be built
than of the larger; and that the stages will be tested individually and
also in combination. Flight testing of any torso part of the entire
4 stage aggregate requires some provision of fairing, otherwise the high
air resistance in the lower atmosphere is likely to seriously impair the
behavior of the missile. This difficulty is proposed to be overcome by

firing the smaller stage units with faired after bodies and the larger stage units with dummy heads. An estimate of the number of units to be built and the various combinations of the four stages proposed to be tested is ventured in the following tables in which capital letters indicate "live" propelled specimens, small letters unpowered dummy specimens.

	live	dummy
Grandmother	3 G	1 g
Mother	18 M	2 m
Daughter	23 D	7 d
Baby	23 B	12 b

A dummy specimen of each may have to be devoted to static tests.

Number of flight tests	12	8	4	4	4	4	2	3	3
Combination	B	Db	DB	Md	MDb	MDB	Gm	GMd	GMDB

The DB and MDb and especially the MDB tests will offer opportunities of carrying telemetering instrumentation aloft which should furnish some of the data needed in the final phase of the project. Incidentally, the MDB aggregate may have immediate applications as a long range missile or weapon. The last item of the list, GMDB represents the firing of the final article from an equatorial location into the orbit. All preceding tests except perhaps MDB are expected to be made on domestic proving grounds with suitably expanded range facilities. The magnitude of this construction and test program is of the order of about 100 V2's, on the basis of corresponding gross weight.

ORM 25-S-1
(REV. 8-43)

PREPARED BY: W. B. Klemperer DOUGLAS AIRCRAFT COMPANY, INC. PAGE: 216

DATE: May 2, 1946 SANTA MONICA PLANT MODEL: #1033

TITLE: PRELIMINARY DESIGN OF SATELLITE VEHICLE REPORT NO. SM-11827

Chapter 15

<u>Facilities required.</u> This program is much more in the nature of an experimental enterprise than a production job. It is therefore imperative that Development, Design, Procurement, Logistics, and Testing be closely coordinated so as to form a continuous loop.

The engineering staff will comprise a large part of the personnel engaged in the project. It will consist of aircraft and rocket power plant engineers, instrument and automatic control experts, and specialists in the sciences and arts of radio, radar, telemetry, trajectory survey, artillery and flight testing.

The manufacturing facilities required will be in the nature of the experimental shop of a large aircraft factory. As there will be little need for quantity production methods, tooling may be best devised by jigging and in many respects by improvisation and adaptation. Provision for static tests of the structures and best provided at the manufacturing plant; static firing test facilities for the rocket engines in existence and under construction will suffice to take care of baby, daughter and mother stages. A suitable test stand to run the grandmother motor would still have to be created. The mother and grandmother units are probably too big to be transported to the flight test site in assembled condition. Provisions for part assembly shipment and reassembly at the test site will have to be made for them. This will comprise elaborate cranes, and scaffold there.

The operation of the baby and daughter can presumably be handled at an existing ordnance test range such as White Sands, N.M. with facilities

FORM 25-S-1
(REV. 8-43)

PREPARED BY: W.B. Klemperer DOUGLAS AIRCRAFT COMPANY, INC.   PAGE: 217

DATE: May 2, 1946 (Corr. 5-28-6)   SANTA MONICA   PLANT   MODEL: #1033

TITLE: PRELIMINARY DESIGN OF SATELLITE VEHICLE   REPORT NO. SM-11827

Chapter   15

now available there. Apparatus necessary to handle the mother stage
which is about as big as two V2's can undoubtedly be developed in stride.
When it comes to Grandmother which are of the order of 10 V2's each as far as
weight and fuel is concerned, a new problem will arise and the question of
whether this phase of the program should be considered at an inland firing
range or located elsewhere, be it at the seashore or at the equatorial orbit emplacement site remains to be investigated.

Shop and field facilities for local erection, assembly, fueling, equipment testing, and observation may have to be largely duplicated at the
equatorial site. The delivery and storage of fuel and liquid oxygen at
the equatorial site may become a sizable enterprise. It may wind up with
the establishment of a local oxygen liquefaction plant either land based
at the site or shipborne on one of the vessels which will have to be assigned to the entire project.

Since the experimental orbit missile emplacement will have to be chosen
within 1 or 2 degrees of latitude, only the following locations are geo-
graphically eligible: Ecuador, N.W. Coast of Brazil around the Amazon
delta, French African Congo Coast; Kenya Colony in Central East Africa;
Straight Settlements and Singapore, Borneo, Celebes, and finally any of
the numerous equatorial Pacific islands between Halmahera and Howland or
Baker. Of these Ecuador and Kenya offer possibilities of accessible moun-
tain sites. Politically, however, it would be preferable to stay in
American controlled territory. For reasons of radio altimetry a site
near an East coast is desirable. Islands for several hundred miles east

DOUGLAS AIRCRAFT COMPANY, INC.

PREPARED BY: W.B. Klemperer

DATE: May 2, 1946 SANTA MONICA PLANT

TITLE: PRELIMINARY DESIGN OF SATELLITE VEHICLE

PAGE: 218

MODEL: #1033

REPORT NO. SM-11827

Chapter 15

of the emplacement site are desirable as fixed observation stations. Most of these considerations point to the archipelago north of New Guinea as the logical region in which to look for islands on which utilizable war installations may be available. Adequate living facilities for the staffs and crews will have to be found or provided at the emplacement site. Rapid communication and transportation facilities between the site and the project headquarters will be a necessity. For the orbital observation and tele-metering some 20 to 50 stations may have to be installed or positioned in a belt around the equator, across the Pacific Ocean, Ecuador, Brazil, Atlantic Ocean, French Congo, Kenya, Indian Ocean and Malaya. All these stations may have to be linked with each other and/or a central director station by a rapid communication system if continuous tracking and telemetering of the satellite missile is to be maintained, and particularly if its return to earth is to be guided.

PREPARED BY: E. Wheaton | DOUGLAS AIRCRAFT COMPANY, INC. | PAGE: 219
DATE: May 2, 1946 | SANTA MONICA PLANT. | MODEL: #1033
TITLE: PRELIMINARY DESIGN OF SATELLITE VEHICLE | REPORT NO. SM-11827

FORM 23-2-1
(REV. 8-45)

Chapter 15

Estimate of Cost - In order to obtain an estimate of the cost of this project, the following assumptions were made:

(a) All development, engineering, and fabrication on the basic vehicle in quantities outlined in the section Program for Building and Testing, estimated in accordance with standard practices in the aircraft industry $50,000,000

(b) All development, engineering, and construction on the following items, arbitrarily assuming to be equal to the work done under (a) above: power plant, propellant pumps, turbines, controls.   50,000,000

(c) All development, engineering, and construction on the following items; instrumentation (payload), establishment of special launching facilities, transportation and analysis of test data assumed to cost   50,000,000

Total Cost of Project. $150,000,000

It should be emphasized that this estimate is very rough, was hurriedly prepared and is without benefit of any experience or actuarial records in this new field. It is possible that due to the large volume of fuel tanks in this vehicle as compared to standard aircraft, this cost estimate is conservative; however, the unknown difficulties which arise in any new field of endeavor would indicate the need of a conservative estimate. One of the important phases of

Chapter 15

the proposed preliminary design phase of the program should be used to further investigate German cost records in the development of the V-2 as a basis for determining costs on this project.

PREPARED BY: H. Klemperer   DOUGLAS AIRCRAFT COMPANY, INC.   PAGE: 221
DATE: May 2, 1946   SANTA MONICA   PLANT   MODEL: #1033
TITLE: PRELIMINARY DESIGN OF SATELLITE VEHICLE   REPORT NO. SM-11827

Chapter   16

16. RESEARCH AND DEVELOPMENT NECESSARY FOR DESIGN

At present sufficient background exists to allow preliminary design studies to begin whenever a decision to proceed is reached. However, a vigorous research program must also be started at once because many data have to be established and estimates verified or improved in order to avoid unnecessary sacrifices of performance. This research program may be divided into several branches of science:

Theory of Structural Enlargement - The engineering steps which lead from established prototype structures to greatly enlarged units are based on certain mechanical concepts of critical design requirements for various parts that perform definite functions. However, no part can be ideally designed for just one function alone; they overlap. In order to take full advantage of past experience it will be necessary to undertake a careful analysis and weight breakdown of sundry past and presently progressing high altitude or long range missile projects, in order to ascertain just exactly to what specifications their essential parts are designed and why. Such analyses of larger units throw light on the degree to which concessions in shape, in margins of safety, in allowances for technological and practical exigencies should be made.

Structural Materials - It seems certain that extensive investigation of the properties of materials will be required, especially the materials to be used for the tank walls, lids, and bottoms. The possibility of using fabric bags deserves consideration. Suitable sealing and insulating materials must be used which at the same time prevent leakage of the fuel and liquid oxygen.

ORM 25-3-1
(REV. 8-43)
PREPARED BY: W. Klemperer   DOUGLAS AIRCRAFT COMPANY, INC.   PAGE: 222

DATE: May 2, 1946   SANTA MONICA   PLANT   MODEL: #1033

TITLE: PRELIMINARY DESIGN OF SATELLITE VEHICLE   REPORT NO. SM-11827

Chapter   16

Tanks - Since much depends upon the weight efficiency of the design of the large fuel and oxygen tank, a serious investigation into the merits of competing tank designs is indicated. Since practical manufacturing, insulating, plumbing and other technical considerations enter into the problem, research will have to go hand in hand with design towards evolving the most advantageous compromise between the various conflicting requirements. Special methods may have to be developed to test specimen structures under simulated acceleration loads.

Stage Separation - Means to separate smoothly the subsequent stage units when the booster unit fuel is exhausted, will have to be developed on the basis of extensive research into the various techniques previously tried and proposed, or yet to be evolved. These studies may include experimental work with reduced and full size dummy missiles prior to application to the full-fledged test rounds. Problems of accurate timing and of minimizing the pitch, yaw, and roll disturbances of the boosted unit will have to be solved. Inter-stage communication of automatic regulation and command signals and its harmless discontinuation upon stage separation will also require research, development and experimentation.

Erection, Assembly, Logistics - The design of the large stage parts of the vehicle will have to meet requirements dictated by considerations of erection and assembly procedures and of shipment of parts and subassemblies to test locations. An investigation into practical handling and operating procedures and into the logistics of the entire project must therefore be completed before the mother and grandmother can be completely designed. No radically new erection methods are believed

Chapter   16

necessary; those well developed in experimental aircraft production
should be reasonably applicable.

Rocket Fuels - The reasons for choosing Liquid Oxygen and Alcohol
as the propellants have been explained in Chapter 6. There are, however,
several other propellant combinations which have a higher specific impulse.
An extensive research program should be initiated to gain information on
motor design criteria, storage and handling problems and the problem of
logistics of such fuels. Several propellant combinations which warrant
immediate consideration are -

   (1) Liquid Hydrogen and Liquid Oxygen

   (2) Hydrazine and Liquid Oxygen

   (3) Methylamine and Liquid Oxygen

   (4) Liquid Ammonia and Liquid Oxygen

   (5) Hydroboron and Liquid Oxygen

   (6) Nitromethanes (as a monofuel)

Any of these after an intensive research program may eventually prove
superior to the Liquid Oxygen-Alcohol combination when all aspects and
technical implications are understood and weighed.

Rocket Motors - Motor research will have to be directed towards improvements in fuel mixing, cooling, combustion chamber shape, nozzle wall
materials, ignition, and pressure control. This research is expected to
lead to reduction of weight and increase in reliability, performance
, and efficiency of motors. Rocket motors are usually tested in test pits
equipped with elaborate laboratory instrumentation and safety devices.
For the largest stage a new test pit surpassing those now existent will

Chapter   16

have to be established.

Rocket Accessories – While no fundamental research problems are foreseen to be involved in the scaling up of present turbines and pumps to the
sizes necessary to feed the larger stages, new problems arise with the
introduction of pump control for thrust throttling. The pump will be
called upon to operate near peak efficiency from full speed down to about
two-thirds speed. How best to accomplish this will have to be determined
by research. No extensive research into plumbing is immediately required,
provided existing standard fuels are used. However, if some new fuel is
contemplated, effect on requirements for insulation, line sizes, valves,
pumps, etc., must be investigated.

Test Stands – The size and cost of the vehicle would preclude the
firing of test rounds in a free flight vehicle for the mere purpose of
testing operational features. Therefore, each stage will have to be
fired in a vertical test stand for proof testing. A test stand suitable
to test run the mother and particularly the grandmother units will be
considerably bigger than the German V-2 test stand. Its flame deflection
and cooling requirements will be severe. The development of such a test
stand with all its service equipment will be a sizable enterprise in
itself and should unfold gradually from test experience gained with the
lesser daughter stages on suitably adapted smaller operational test stands
now available. Among things requiring proof testing and measuring, may
be listed: the length of life of the jet vanes, the magnitude of malalign-
ment of the jet, time lag of controls, magnitude of the thrust and of
the control forces, functioning of all pumping machinery, etc.

Chapter 16

Problems in Gasdynamics - A vehicle of the type discussed in this report will have to travel through air varying in density from sea level values of .0024 slugs per cubic foot to those of the order of 10^{-13} at 200 miles altitude. The atmospheric conditions encountered thus vary from the standard sea-level values to those of an extremely rarefied gas. The physical phenomena, consequently, vary considerably, and so, of course, do the methods of evaluation. It is convenient to distinguish the regions encountered by the ratio of the characteristic length parameter of the body to the mean free path of the gas molecules. It is also convenient to introduce, in addition to some suitable dimension l of the vehicle, another length, the boundary layer thickness δ which in terms of the velocity U and the kinematic viscosity ν is related to l by

$$\delta^2 \approx \frac{\nu l}{U} \tag{1}$$

The quantity ν is found from elementary kinetics of gases as

$$\nu \approx \Lambda\, \bar{c}, \tag{2}$$

where $\bar{c}$ is the mean molecular velocity and Λ the mean free path. Following Tsien[1] we may now immediately indicate regimes of flow by comparing with Λ and l . δ

(a) $\Lambda << \delta$

(b) $\Lambda \sim \delta$

(c) $\delta < \Lambda < l$

(d) $\Lambda >> l$

[1] H. S. Tsien: Symposium on High Speed Aerodynamics. Cal. Inst. of Tech., March 1946. (Unpublished)

Region (a) is the realm of aerodynamics, in the usual sense. (b) is the region in which slip phenomena occur in the velocity distribution near a boundary. Temperature discontinuities also occur at the wall. Region (c) is the least understood case since in this region deviations from the Maxwellian velocity distribution play a large part. (d) is sometimes called the "Knudsen" region in which pure gas-kinetic methods apply.

Speaking in more aerodynamical terms, it is possible to distinguish the four regions in terms of Mach number, M, and Reynolds number, R, where

$$R = \frac{U l}{\nu}, \quad \text{and}$$

$$M = \frac{U}{a} \approx \frac{U}{c}$$

where a denotes the velocity of sound. Then from equations (1) and (2) it follows that

$$\frac{\delta}{l} \approx \frac{1}{\sqrt{R}} \approx \sqrt{\frac{\Lambda}{l} \times \frac{1}{M}},$$

and we therefore have

$$\frac{\Lambda}{\delta} \approx \frac{M}{\sqrt{R}}. \tag{3}$$

A given problem, therefore, can be characterized by the mean free path and boundary layer thickness, or by the Mach number and the Reynolds Number. For example, orbital conditions at 100 miles altitude correspond to $M \sim 17$ and $R \sim 60$. Hence, $\Lambda/\delta \approx 2$ and thus the problem is in the slip region.

The research problems encountered in these various realms are briefly listed in the following paragraphs. First the case of very large mean free path is discussed. Then the problems of larger densities are taken up. The problems of air resistance and heat transfer in rarefied gases which are encountered in the present analysis differ from most problems of this kind previously studied. The former, such as those related to the oil drop experiment, etc., concerned problems in which the mean velocity was small compared with the molecular velocities. The problem here involves velocities which are large compared with the random molecular velocities. Hence the problem of momentum and energy transfer encountered here is similar to the interaction of a molecular beam of nearly uniform velocity and direction interacting with a solid surface.

Experiment and theory are both not too well advanced in considerations of this type. It seems highly desirable to initiate a research program. This work should start with a general review of what is already known, both experimentally and theoretically, and then proceed to fill the gaps in the knowledge. To mention a few assumptions which have not yet been completely substantiated, we have:

1. The reflection of molecules from a surface is generally assumed to be diffuse. The molecules are assumed to leave in random directions.

2. The departing molecules are assumed to have velocities related to the temperature of the wall by means of an empirical "accommodation" coefficient A, which varies between zero and one. $A = 0$ corresponds to the case of elastic impact, $A = 1$ to the case where the temperature of the molecules of the wall is equal to the wall temperature.

Both of these assumptions affect directly the evaluation of drag and heat transfer. For the present case of high velocities a complication

Chapter   16

arises due to the fact that the molecules will not leave the wall, but will form a surface layer. This effect cannot yet be considered in the calculations and further research is needed.

Conditions at somewhat higher densities are still more complicated. In regimes where the mean free path is comparable with the boundary layer thickness but not large compared to the body dimensions, complications arise due to the fact that deviations from the steady state have to be considered. The necessary additional terms in the equations of motion due to the deviations from a steady state can be calculated from the work of Chapman and Enskog. Tsien has pointed out that the additional terms in the Navier-Stokes equations involve third order differential quotients and hence a new boundary condition has to be added. The exact form of this boundary condition is at present unknown. The formulation of such a condition will again involve knowledge of the interaction between the gas and the solid wall.

The gasdynamical range appears to be the best understood at present and considerable research here is already under way. New problems related to designs such as the present one are, for example, the non-stationary heat transfer problems which have to be considered for ascent and landing. Questions like the drag curve at high values of M, stability in passing through the sonic velocity, etc., are obviously of great importance. However, there is little difference in the gasdynamical range between problems arising in ordinary missile design and problems of the design of space vehicles. Of course, detailed design problems will differ.

It should be emphasized here that aerodynamical research related to a space vehicle will have to make use not only of extensive investigation

Chapter 16

in wind tunnels, both supersonic and hypersonic, but will also have to use many tools of modern experimental physics. The molecular beam methods, for example, will be one of the required tools. It is clear that in many cases the research has to be analytical and not experimental. In low speed aerodynamical research the method of model tests simulating actual flight conditions is extensively used. This type of test becomes exceedingly difficult and in fact impossible for vehicles travelling at great speeds. This is most easily seen in the following example. The ratio of stagnation temperature T_0 to free-stream temperature T for a body travelling at a Mach number M is

$$\frac{T_0}{T} = 1 + \frac{\gamma - 1}{2} M^2, \quad \text{where}$$

$$\gamma = \frac{C_p}{C_v}$$

Hence for $M = 10$ and $\gamma = 1.4$, $T_0/T = 21$. To produce the Mach number 10 in a wind tunnel, an adiabatic expansion process is used and the temperature drops from the value T_0 in the supply section to $T_0/21$ in the test section. In free flight, on the other hand, the free-air temperature T is given, and the temperature $T_0 = 21\ T$ is produced by compression due to the fast moving vehicle. It is evidently very difficult indeed to match both M and T in the wind tunnel as compared with flight, since this would require a temperature in the supply section of the wind tunnel 21 times larger than the ambient temperature in flight. Research must become more analytical, and a close cooperation between experiment and theory and also between aerodynamics and physics is an absolute necessity.

Chapter 16

Aerology - Knowledge of the properties of the upper atmosphere above the stratosphere is expected to be greatly advanced during the present year as the V-2 firing program and the Hermes, Bumblebee, WAC and Nike projects progress. Arrangements will have to be made to cooperate with these programs towards securing some of the most important aerological data at an early stage. The factors of primary interest are: air temperature, atmospheric composition ionization, radiation and wind. As the satellite project itself develops, it will gradually furnish its own means of pushing exploration into the realm of higher altitude and speeds to determine the laws which govern the behavior of moving bodies there. Some of the incomplete missile aggregates, notably the mother - daughter - baby stage test rounds when fired on ballestic trajectories over hundreds or a few thousand miles range should furnish excellent opportunities of penetrating these virgin regions. Suitable instrumentation with which these and other test missiles can be equipped should be developed as the project gets under way.

Jet Control Rudders.- The jet rudders for control of this vehicle while under thrust presents a major problem in that they will be required to stay in the jet about four times longer than any that have been used to date. The natural Graphite vanes on the German V-2 lasted about 30 seconds before erosion rendered them useless for control. It therefore will be necessary to enter into a research program to determine what material or combination of materials will be able to withstand the high gas velocities of the hot jet sufficiently long. A passable solution may be in the almost unexplored powdered metals and sintered ceramics.

Another approach may be to have a number of vanes set at a given angle to the jet exhaust and as control is required the vanes feed into the jet.

Chapter 16

As the vanes erode away the control would feed more and more of the vanes into the jet. The solution of the problem lies at the end of a research program.

Servo System - A considerable amount of research will have to be directed towards the development of a suitable servo motor system to actuate the jet vane. For instance, the relative merits of electric, pneumatic, hydraulic systems have to be investigated for each stage. Means to balance the vanes in order to keep their torque and power demands in bounds will have to be studied.

The control of the servo motors by means of an automatic program pilot system, its stability and freedom from undesirable hunting will come in for extensive research and development. It will require gyroscopic and accelerometric response elements which themselves will have to be specially developed even though their fundamental principles have already been successful in the V-2 missile. To these will be added a radio altimeter, the development of which will constitute a program beginning with research into the special functional requirement of such a device, when working, at unprecedented altitudes and flight speeds and furnishing input signals to an automatic control regulator.

Guide Beam and Command System.- Inasmuch as it is anticipated that provision will have to be made to signal corrective guide commands to the missile to ensure its precise entry into the desired orbit, a good deal of research and experimentation with various competing guide beam and guide command systems must be instituted in order to evolve a satisfactory technique for the satellite. Antenna systems will have to be developed for the satellite missile in all its stages. Research will be directed

Chapter 16

towards determining the most suitable wave lengths for communication with the missile and towards developing suitable tubes for them, if none are available. Preliminary trials of proposed ground installations, as well as air borne radio altimeters, command receivers and transponders, should be arranged in conjunction with other missile firing projects in progress, and subsequently, with test missiles specially adapted for this purpose. During the later stages such tests can undoubtedly be combined with flight tests of the daughter stages of the satellite project itself. The command system eventually charged with bringing the missile back to earth will also constitute an object for research and development. The electric power supply for the missile borne electric equipment will also have to be developed to suit the unusual conditions prevailing on the orbit in an environment without air and gravity.

Simulator.- In order to ensure that the complex regulator loop system will work smoothly and without undesirable hunting it seems imperative that a so-called simulator be built and operated over numerous test runs. Such simulators have proved themselves invaluable in conjunction with many projects, viz Roc, Azon, Bumblebee, Hermes, Nike, V-2, FX, Henschel, etc. The development of an electrical simulator for the satellite missile can be undertaken in such a manner that it can be quickly adapted to represent many varieties of control systems with different response characteristics, lags, coordination, overrides etc. The representation will include the automatic control functions as well as the beam commands when superimposed, and it will be made to take care of the gradual changes of missile weight, moment of inertia, aerodynamic reactions, thrust, etc. during each simulated flight.

Attitude Control in Orbit. - Reaction flywheels or recoils which are

Chapter 16

being considered for control of the missile's attitude or orientation with respect to its flight path, will require research and development to ensure that they function as desired under the peculiar environment of the satellite and that their control is sufficiently forcible. Simulator techniques may also be helpful here.

Telemetry, Trajectory Survey and Communication. - Telemetry from guided missile to ground stations is still in its infancy. Great strides towards its realization are expected to be made as the presently active projects, notably the V-2 program, Hermes, Bumblebee, etc., progress. Close coordination with these activities will be mandatory. Later on, special firing of some missiles for the benefit of the perfection of satellite telemetering systems may be indicated. The same is true of the evolution of trajectory survey by photo and kine-theodolite and radar tracking gear. Eventually the entire communication system, which will be required in the actual satellite operation, will have to be developed, tried out and practiced. The development of the instrumentation for the "payload" will require specialized research along the lines of the various branches of science to which the data to be measured belong, notably, meteorological instruments, cosmic ray and ionization measurements, temperature and pressure measurements, radiation measurements, spectroscopy, photography, television, etc.

Digest of Existing Literature. - The scientists of many nations have written more or less technical or speculative articles on many objects pertinent to the satellite project. German scientists and engineers, in particular, have produced numerous reports covering almost every phase of their research and development work in conjunction with their missile projects, several of which have been technically radical advancements of the

Chapter 16

art. These reports are physically available in this country. They are, however, scattered between several agencies (Army Air Forces, Army Ordnance, Navy BuAer, Navy Ordnance and various other institutes.) They are in the process of being indexed, screened, abstracted and microfilmed. Some few have been translated. Translations are being made by specialists scattered throughout the nation-wide galaxy of activities in the field of guided missiles. One of the urgent tasks of the research agency of the satellite project would be to scan this wealth of literature, gather copies of translations of significant reports now available or in progress, obtain microfilms of others and organize the translations of those that have an immediate bearing on the yet unsolved or uncertain problems of the satellite project. In some instances it may be possible to secure from the military authorities the assistance of German authors or specialists to expedite the jobs of translating or abstracting the material.

Interrogations - In many instances discrepancies or conflicts appear in the literature between data, computations, theories and objectives of some phase or other of the German activities. Ad hoc questioning of German personnel held available in this country has already done much to clarify some of the problems involved. Even though these people have been interrogated many times it seems that as the digest of the literature and the transcript progresses and broadens, new questions arise continuously. It is therefore believed advisable to comprise in the research activities of the satellite project some set-up whereby discussions of such new questions can be quickly arranged between the German personnel and the American specialists on the problems.

Coordination with other projects - As has already been mentioned

Chapter 16

under various headings before, part of the necessary research activity will comprise close coordination with other high altitude and long range missile projects now actively being pressed forward in this country, and - if possible - abroad.

Continuity - Some of the research problems are of a fundamental nature and have a bearing on the first moves of the design staff. These will have to be tackled immediately. Others are directed towards results needed at various later stages of the program. These can be scheduled consecutively. A certain amount of overlap and continuity of research and development will therefore become necessary. Actually research and development will have to be continued to the very end of the project, the actual firing and navigation of the satellite missile is more a research task than a production job.

Chapter 17

17. CONCLUSION

In the preceding chapters, we have critically examined the possibility of designing a man-made satellite. This examination has been made within the strict limits of practical engineering analysis. We have found that modern technology has advanced to a point where it now appears feasible to undertake the design of such a satellite.

The magnitude of the task of establishing the first satellite vehicle in its orbit is impressively large. However, our analysis has shown that as experience is gained in this new field, the reduction in the magnitude of the task will be equally impressive.

#1033

Appendix A

A. THE UPPER ATMOSPHERE

In evaluating the performance of a very high altitude vehicle, such as that described in this report, it becomes necessary to have values for the physical properties of the upper atmosphere at extremely high altitudes, which heretofore were of little interest to the aeronautical engineer. Conditions in these high altitude regions have received some attention, both theoretical and experimental, in the past 20 or 25 years by a relatively small number of investigators. However, the present knowledge of the physical state of the upper atmosphere is far from complete, and as will become apparent in the course of the discussion, at the high levels there is quite some differences of opinion as to what the conditions are; at still higher levels there are practically no data or opinions available at all. In short, the knowledge of the atmosphere becomes more and more uncertain and speculative with increasing altitude. The knowledge which is available concerning the upper atmosphere is based essentially on the results of observations of meteors, the spectrum and height of the aurora, the behavior of radio waves, the anamalous propagation of sound, and the ionization of the atmosphere by solar radiation.

In general, workers in the field appear to be in fair agreement as to the atmospheric properties from sea level up to 60 miles altitude. Above this altitude the knowledge and agreement is much less definite. It should be mentioned at this point that since all of

FORM 25-S-1 (REV. 5-45)

PREPARED BY: G.Grimminger
DATE: May 2, 1946
TITLE: PRELIMINARY DESIGN OF SATELLITE VEHICLE

DOUGLAS AIRCRAFT COMPANY, INC.
SANTA MONICA PLANT

PAGE: 2A
MODEL: #1033
REPORT NO. SM-11827

Appendix **A**

the values given in the literature are based on the metric system, temperature in $^\circ K$, altitude in km., the values which will be quoted here will be in terms of the same system. For the convenience of the reader, tables 4 and 5 for converting values of temperature and altitude to the English system, $^\circ R$ and ft., are given at the end of the text. The values finally adopted to represent the atmosphere will be presented in the engineering system of units.

Since the pressure at any altitude depends on the vertical distribution of temperature, and since the density depends on both temperature and pressure, it is evident that, of any of the atmospheric properties, the temperature is the most fundamental and important one to be considered. The discussion which follows is undertaken with this point in mind.

For the purpose of discussion, the atmosphere is usually divided into three main regions. The atmosphere from sea level to about 10 km. is referred to as the troposphere and that from 10 km. to 20 km. as the stratosphere. The region above 20 km., extending outward to interplanetary space (or however far outward the atmosphere may be considered to extend) is referred to as the upper atmosphere. As pointed out by Penndorf[1], measurements of auroral heights indicate the presence of atmosphere up to heights of 1000-1200 km. The average

(1) Penndorf, R.: Die Zusammensetzung der Luft in der hohen Atmosphäre. Meteorologische Zeitschrift, vol. 55; p. 30, 1938.

Appendix A

conditions in the troposphere and stratosphere are well known and form the basis for the standard atmosphere used in aeronautics, as given by Diehl[2].

The atmosphere above about 80 km. is strongly ionized and hence this region of the upper atmosphere from 80 km. outward is known as the ionosphere. The ionosphere is of fundamental importance in radio-wave propagation since it is owing to the reflection of these waves by the ionosphere that long distance radio communication is possible. It seems to be established that the ionization, and therefore the conductivity of the ionosphere, is caused by the ultra-violet solar radiation. Whether the particles responsible for the conductivity are ions or electrons has not been definitely established, especially for the lower levels of the ionosphere. A survey of the facts and theories of the ionosphere has been given by Mimno[3].

The ionosphere itself is divided into three main regions or layers. According to Berkner[4] the lower of these regions known as

(2) Diehl, W.S.: Standard Atmosphere - Tables and Data. NACA Technical Report No. 218; 1925 and 1940.

(3) Mimno, H.R.: The Physics of The Ionosphere, Reviews of Modern Physics, vol. 9, No. 1; Jan., 1937.

(4) Berkner, L.V.: Physics of The Earth - VIII, Terrestrial Magnetism and Electricity, McGraw-Hill; p. 451, 1939.

Appendix A

the E-region is moderately ionized and is situated in the vicinity of the 100 km. level. The next higher layer, the F_1-region, is more strongly ionized and is situated at about 210 km. Still higher and still more strongly ionized is the F_2-layer at about 300 km. Most of the present knowledge of the upper atmosphere is based on the study of these three regions plus some figures which have been deduced from meteor studies, the propagation of sound, and the spectrum of the aurora.

Since the aim of this study is to arrive at working values for a so-called standard upper atmosphere, the various data which are available, either experimental or theoretical, will be presented and from them, what appear to be the most reasonable deductions will serve as a basis for the final values to be adopted for use in the study of the performance of the satellite vehicle.

Gutenberg[5] has recently given values for the atmosphere from sea level to 100 km. These are presented here in Table 1, which has been copied from the paper by Gutenberg.

---

(5) Gutenberg, B.: The Physical Properties, Pressure, Temperature and Composition of the Upper Atmosphere. Aeronautical Symposium at the California Institute of Technology; March 1946.

TABLE 1.

Typical Data for the Atmosphere in Winter

(a) Northern Germany, (b) Southern California
(From Gutenberg, Ref. 5)

Altitude km	Temperature °C (a)	Temperature °C (b)	Density $\frac{g}{m^3}$ (a)	Density $\frac{g}{m^3}$ (b)	Sound Velocity $\frac{m}{sec}$ (a)	Sound Velocity $\frac{m}{sec}$ (b)	Mean Free Path, cm	Prevailing Wind Vel. $\frac{m}{sec}$	Prevailing Wind Dir. from	Pressure, millibars (a)	Pressure, millibars (b)
0	3	10	1280	1250	334	338	6.1×10^{-6}	5+ 5−	W	1015	1015
4	−17	−5	825	802	322	329	9.5×10^{-6}	increasing		607	618
8	−45	−35	525	528	304	310	1.5×10^{-5}	to tropopause		344	361
12	−52	−65	294	326	299	291	2.3×10^{-5}	20+ 20−	W	187	195
16	−50	−80	157	180	300	279	4.2×10^{-5}	decreasing		101	99
20	−48	−70	85	84	301	286	9×10^{-5}	variable		55	49
24	−44	−55	46	41	304	297	2×10^{-4}	variable		30	26
28	−20	−20	23	19	320	320	4×10^{-4}	increasing		17	14
32	+20	+10	12	10	345	339	7×10^{-4}	30+ 30−	E	10.2	8.7
40	+65	+35	5	4	370	353	.002	100+ 100−	E	4.4	3.4
50	+80	+70	1.6	1.2	376	372	.006	100+ 100−	E	1.6	1.2
60	+80	+75	0.5?		372	375	.02	150+ 150−	E	0.6	0.4
80	−10?		.08?		326?		0.1	150+ 150−	E	.07?	.05?
100	−150?		.006?		415?		1	changing to Westerly		.007?	.007?

TABLE 1
APPENDIX A

Hulbert,[6] another investigator in problems of the upper atmosphere, adopts the temperature distribution shown in Table 2 to represent the atmosphere up to 220 km.

Table 2

Temperature of the Day Atmosphere
(According to Hulbert, Ref. 6)

Altitude, km.	0	10	20	30	40	60	80	100	200	220
Temperature, °K	287	220	225	230	240	260	320	360	360	360

Fig. 1 copied from a paper by Martyn and Pulley[7] also gives information, actual and inferential, concerning the average vertical temperature disbribution. Above 100 km. it will be noticed that the temperature distribution has been extrapolated up to 300 km. as shown by the broken lines.

Fig. 2, copied from the paper by Penndorf, loc.cit. an equally reliable investigator, gives a somewhat different curve for the probable vertical temperature distribution above 100 km. The fact that Goetz[8], another well known worker in the field, uses Penndorf's curve may perhaps lend added weight to the values shown in Fig. 2. It will be noticed that the composition of the atmosphere, according to Penndorf, is indicated by the curve on the left hand side of Fig. 2.

---

(6) Hulbert, E.O.: Physics of The Earth - VIII, Terrestrial Magnetisum and Electricity, McGraw Hill; p.493, 1939.
(7) Martyn, D.F. and Pulley, O.O.; The Temperatures and Constituents of the Upper Atmosphere. Proc. Roy.Soc., A154; p.482,1936
(8) Goetz, F.W.P.: Ergebnisse Der Kosmischen Physik-Band III - Physik Der Atmosphäre, Leipzig, Akad.Verlog.; p.314,1938.

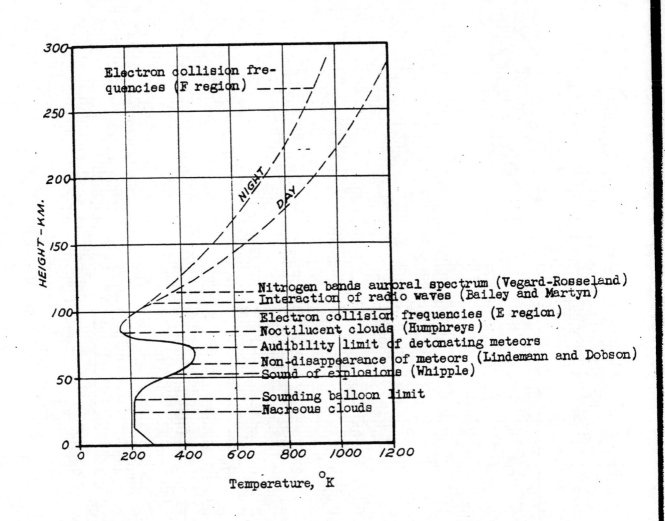

Fig. 1 Temperature Distribution in the Upper Atmosphere According to
Martyn and Pulley (Ref. 7).

Analysis PRELIMINARY DESIGN OF SATELLITE VEHICLE
Prepared by G. Grimminger
Date May 2, 1946

DOUGLAS AIRCRAFT COMPANY, INC.

Page 8A
Model #1033
Plant
Report No. SM-11827

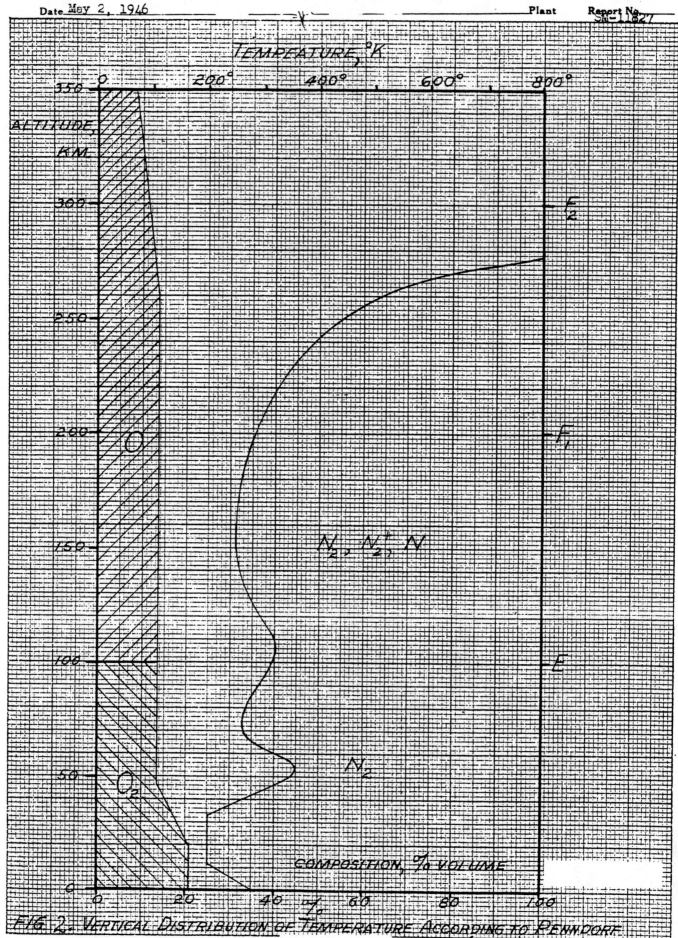

FIG. 2. VERTICAL DISTRIBUTION OF TEMPERATURE ACCORDING TO PENNDORF

Appendix A

The data presented thus far in Tables 1 and 2 and in Figs. 1 and 2, plus the standard atmosphere data of ref. 2 may safely be said to represent the present state of knowledge concerning the vertical distribution of the temperature of atmosphere up to altitudes of about 300 km. It is worth noting that none of the data extend beyond this level. Thus there is a relatively large region extending from 300 km. up to 1200 km. and higher in which, although it comprises only a small part of the atmospheric mass, the temperature conditions are more or less unknown.

That there is not even complete agreement in the region from 60 to 300 km. is immediately evident from a comparison of Table 1, Table 2, Fig. 1 and Fig. 2.

The data presented here, plus the results of other investigators which may be used to throw additional light on the problem, will now be used in order to arrive at the some final and definite values of the probable average vertical temperature distribution which may be adopted to represent a standard upper atmosphere.

From sea level up to 20 km. the atmospheric temperature has been determined by a great many direct measurements (sounding balloons) and the average conditions are well represented up to 65,000 ft. by the values given by Diehl (loc. cit.) for the NACA standard atmosphere. Owing to the universal acceptance and widespread use of the NACA standard atmosphere, it will be adopted as a representation of the atmosphere up to 65,000 ft.

Above the stratosphere, all of the data except that of Hulburt indicate a maximum in the temperature curve at 50-60 km. The existence of this maximum is fairly well established by the work of

Appendix A

(9)   (10)
F. J. W. Whipple and Duckert on the anomalous propagation of sound, the
(11)   (12)
work of Gowan and Dobson on ozone, by the more recent investigation of
(13)
F. L. Whipple based on meteor observations, and also in a recent paper
(14)
by Gowan. The temperatures at 50-60 km. given by Gutenberg agree essen-

tially with the most recent results of F. L. Whipple and Gowan and these

values will be adopted. The temperature at this level given in Fig. 1

appear to be too high.

There is some evidence for a temperature minimum at about the

80 km. level but the value given in Fig. 1 appears much too low, see
(14A)
F. L. Whipple, ref. 13 and Martyn, and we again adopt the more conserva-

tive value given by Gutenberg which agrees with that shown by Penndorf.

(9) F. J. W. Whipple: Quart. Jour. Roy. Met. Soc. Vol. 60; p. 80, 1934.

(10) Duckert, P.: Gerlands Beiträge Zur Geophysik, Supplement 1, p. 280, 1931

(11) Gowan, E. H.: Proc. Roy. Soc., Vol. A128, p. 531, 1930.

(12) Dobson, G. M. B.: Proc. Roy. Soc., Vol. A129; p. 411, 1930.

(13) Whipple, F. L.: Meteors and the Earth's Upper Atmosphere, Reviews of
Modern Physics, Vol. 18, No. 4, p. 246, 1943.

(14) Gowan, E. H.: Note on Ozonosphere Temperatures. Aeronautical
Symposium at the California Institute of Technology; March, 1946.

Appendix **A**

Near and above the 100 km. level another rise in temperature
is required by the results of the studies of the auroral spectrum by
Vegard[15] and Rosseland and Steensholt[16] which give a temperature
about the same as that of Hulbert and which is about midway between the
values given by Martyn and Pulley and that given by Gutenberg. We
therefore adopt the value in Table 2 at the 100 km. level.

Above 100 km. the variation of temperature becomes much less
definite although it is generally agreed that high temperatures must
exist in the F_2-region at 250 or 300 km. Godfrey and Price [17] have
shown, on the basis of radiation equilibruim, that the highest <u>possible</u>
equilibrium day-time temperature in the F_2-region is about $3300°K$,
and that the actual equilibrium temperature may have any value between
this figure and $230°K$. However, these authors have shown that the
existence of high temperatures of the order of $1000°K$ or more is a
necessary consequence of the presence of appreciable oxygen at these
levels.

Martyn and Pulley, loc. cit., also agree that the temperature of the
F-region must be of the order of $1000°K$, and Martyn (ref. 14a) more
recently states that the high temperatures originally found to exist in
the F_2-region as a result of electron collision frequency measurements
is confirmed by Fuchs and by Appleton from measurements of the thickness
of this region. The weight of evidence in favor of high temperatures in
the F_2-region is very considerable.

(15) Vegard, L.: Geophysiske Pub. Oslo, No. 9, 1932
(16) Rosseland, S and Steensholt, G.: Univ. Obs. Oslo; Publ. No.7, 1933
(14a) Martyn, D.F.: The Upper Atmosphere. Quart. Jour. Roy. Met. Soc.
vol. 65; p. 329, 1939
(17) Godfrey, G.H. and Price, W.L.: Proc. Roy Soc., vol. A163; p. 237,
1937

Appendix A

Although there is general agreement on the existence of high temperatures at around 300 km. there is certainly nothing which definitely fixes the shape of the temperature curve between 100 and 300km. On the one hand we have the extrapolated curve of Martyn and Pulley, Fig. 1, which indicates rapidly increasing temperature starting at 80km. and continuing to a maximum at around 300km. On the other hand there is the curve of Penndorf showing practically an isothermal condition from 100 to 200km. and then a very rapid increase in the F_2-region above 200km. As far as the computation of pressure and density is concerned, these two curves would lead to considerably different results. The use of Penndorfs curve would lead to low values of pressure and density at high altitudes, while the use of the curve of Martyn and Pulley would lead to relatively high values for these qualities.

Thus, although Figs. 1 and 2 are in agreement as to the value of the high temperature at 300km., they represent the two extremes by which this temperature is reached starting from 100km. As a reasonable compromise for the probable temperature variation from 100 to 300km., it has been decided to adopt a temperature variation in this region which is an average of these two extremes.

Above 300 km. there are hardly any data which would serve to extend the temperature curve to higher altitudes. We do know that the thickness of the total F-region (comprising both the F_1 and F_2-region) is estimated to be of the order of 200km. thick (Martyn and Pulley, loc. cit. p.469) so that above the 300 km. level one might expect the temperatures to decrease fairly rapidly. Above this level, the only figure which has come to the attention of the writer is a value

of 70°C at 1000km. which is quoted by Rosseland[18] and is due to Vegard.

(18) Rosseland, S.: Theoretical Astrophysics. Oxford University Press; p.237

Appendix A

This temperature is based on high auraral observations and was computed on the assumption of thermal equilibrium. That thermal equilibrium does not obtain under the conditions existing at these altitudes is generally recognized. However, for lack of anything better, the value of 70°C at 1000 km. will be adopted as giving some indication, at least, of the probable order of magnitude.

The final values adopted (as described above) to represent the vertical distribution of temperatures in the upper atmosphere are indicated by the temperature curve presented in Fig. 3.

The composition of the upper atmosphere will now be briefly considered. The composition of the air in the troposphere (sea level to 10-20km) as given by Paneth [19] is shown by Table 3.

From the Table it is seen that N_2 and O_2 account for 99 percent of the composition, by volume, of the lower atmosphere. According to Penndorf (ref. 1, p. 31) and to Chapman [21], the results of auroral spectroscopy indicate that even from 100km to 1000km., oxygen and nitrogen are still the main constituents of the atmosphere. Some twenty years ago, Chapman and Milne [20], it was thought that because of their low molecular weights, either hydrogen or helium must be the main constituent in the high atmosphere. However, it is now the

(19) Paneth, F.A.: Composition of the Upper Atmosphere - Direct Chemical Investigation. Quart. Journ. Roy. Met. Soc., vol. 65; pp. 304-310, 1939

(20) Chapman, S. & Milne, E₂A.: The Composition Ionisation and Viscosity of the Atmosphere at Great Heights. Quar. Journ. Roy Met. Soc⁺, Vol. 46; p. 379, 1929

(21) Chapman, S.: The Upper Atmosphere Quar. Journ. Roy Met. Soc., Vol. 65; p. 304, 1939.

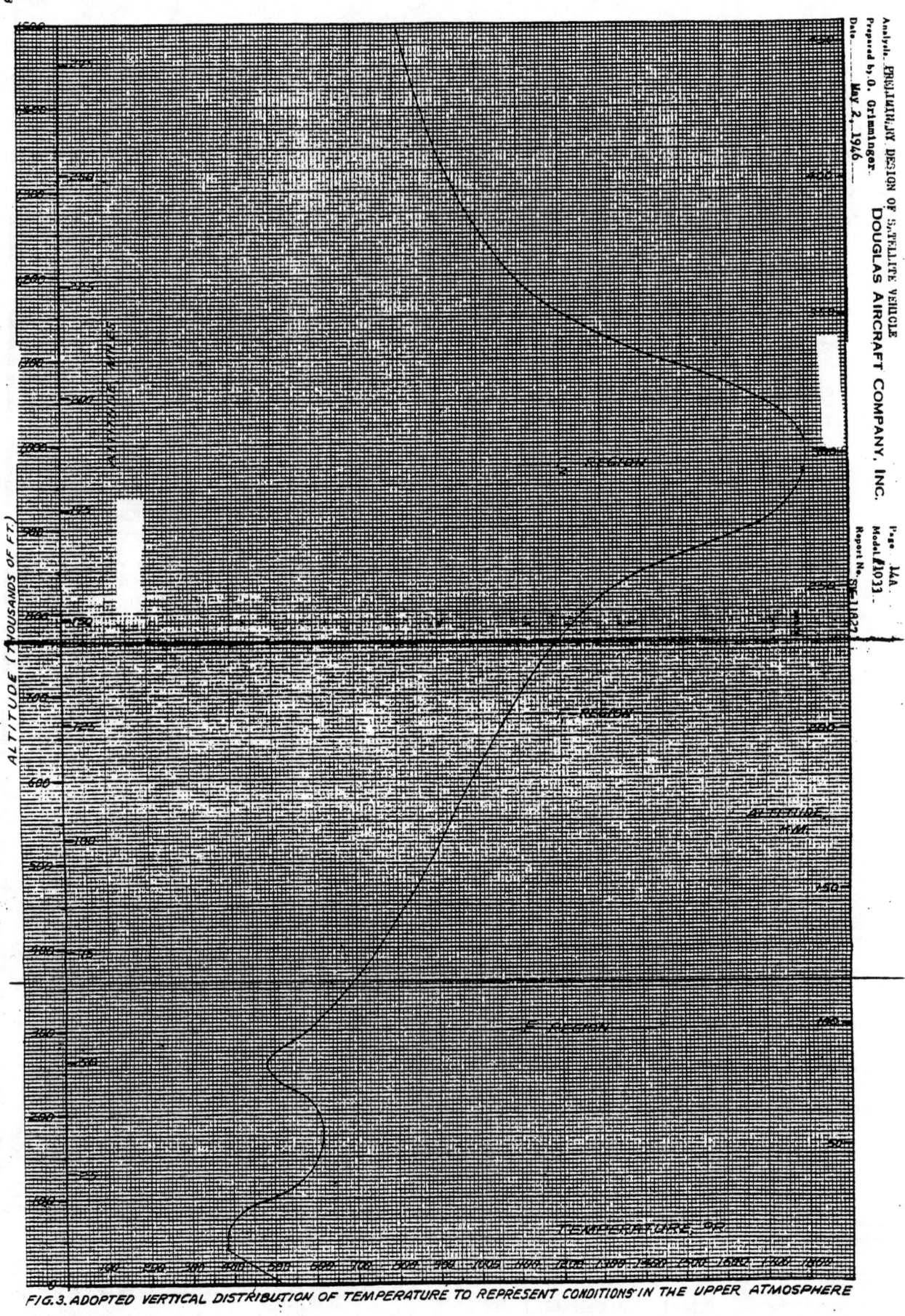

FIG. 3. ADOPTED VERTICAL DISTRIBUTION OF TEMPERATURE TO REPRESENT CONDITIONS IN THE UPPER ATMOSPHERE

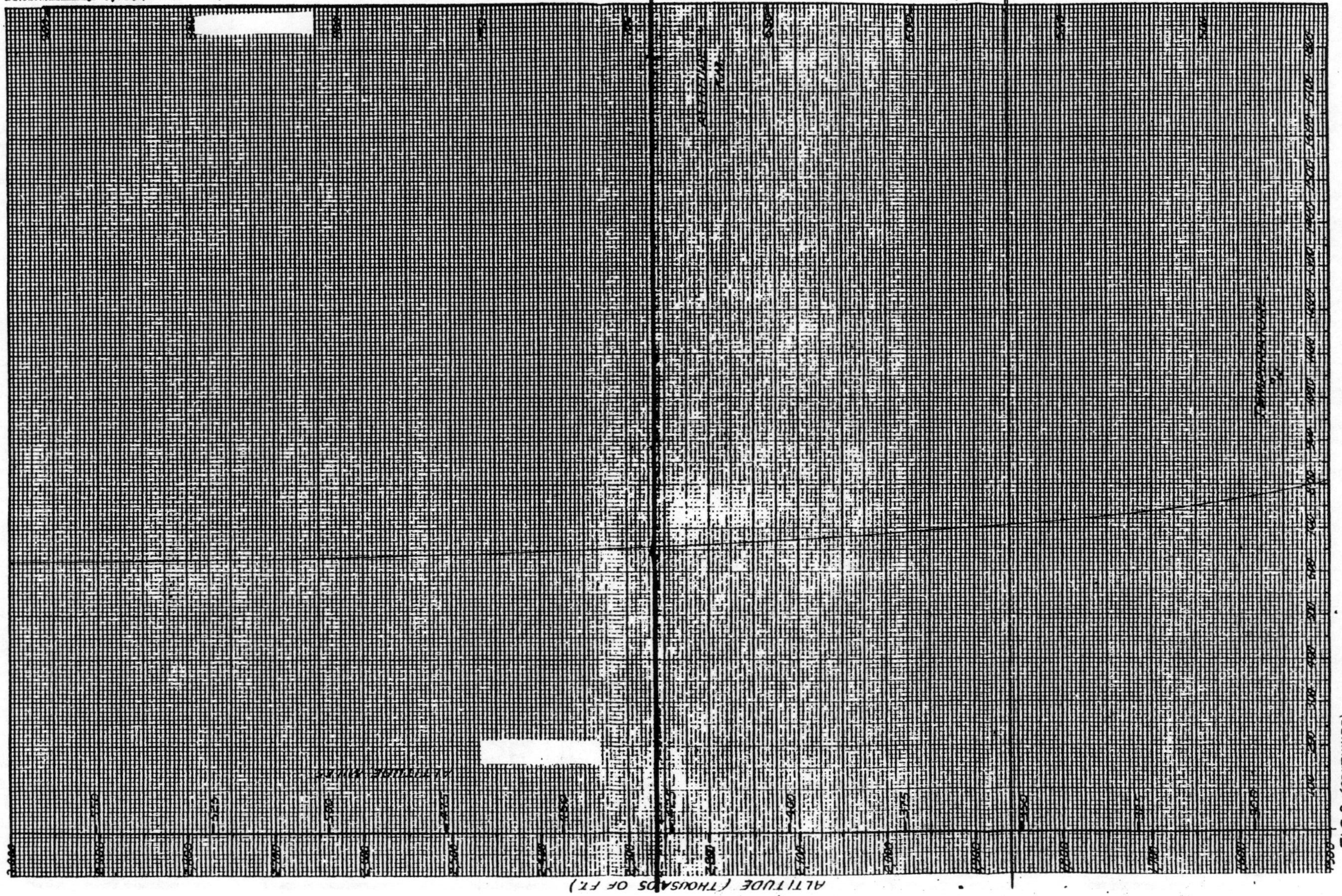

FIG.3-(CONTINUED)

Analysis PRELIMINARY DESIGN OF SATELLITE VEHICLE Page 15A

Prepared by G. Ortmeuiger DOUGLAS AIRCRAFT COMPANY, INC. Model #1033

Date May 2, 1946 Report No. SM-11827

ALTITUDE (THOUSANDS OF FT.)

Appendix A

concensus of opinion that the upper atmosphere is a nitrogen-oxygen

atmosphere, although the presence of hydrogen and helium has not been

absolutely disproved, see Lindenmann, p. 331 of ref. (21).   It will

be assumed here that the upper atmosphere is a nitrogen-oxygen atmosphere.

TABLE 3

COMPOSITION OF TROPOSPHERIC AIR. AFTER PANETH.

Gas	Formula	Volume %	Molecular Wt. (O=16.000)	Density (Air=1)
Nitrogen	N_2	78.09	28.016	0.967
Oxygen	O_2	20.95	32.000	1.105
Argon	Ar	0.93	39.944	1.379
Carbon Dioxide	CO_2	0.03	44.00	1.529
Neon	Ne	$1.8 \cdot 10^{-3}$	20.183	0.695
Helium	He	5.24×10^{-4}	4.002	0.138
Krypton	Kr	1.10×10^{-4}	83.7	2.868
Hydrogen	H_2	5.10×10^{-5}	2.016	0.0695
Xenon	X	8.10×10^{-6}	131.3	4.525

Chapman, ref. 21; and also Penndorf, Fig. 2, agree that the

molecular oxygen O_2 must begin to undergo dissociation into atomic

oxygen O beginning at 100-150km., and that the molecular nitrogen N_2

must undergo dissociation into atomic nitrogen at higher levels.

In this report the following values will be adopted for the composition

of the atmosphere.

Appendix A

Altitude Range	Composition	Molecular Weight of Mixture
0-150KM	21% O_2 and 78% N_2	28.7
150-500km.	12% O and 88% N_2	26.5
500 km. and higher	10% O and 90% N	14.2

Having adopted the temperature distribution shown in Fig. 3 to represent the probable average conditions in the upper atmosphere, the corresponding pressures are determined by use of the hypsometric equation (see Humphreys[22] or Diehl, ref.2), which is used here in the form

$$\text{Log}_{10}\, p_f = \log_{10} p_1 - 0.000281\, m \left[1 - \frac{2h_f}{R} \right] \frac{h_f - h_1}{T_m}, \quad \frac{lb}{sq.ft.}, \quad - (1)$$

where h_1 = altitude of lower level, ft.

h_f = altitude of upper level, ft.

p_1 = pressure at the lower level, $\dfrac{lb}{sq.\,ft.}$

p_f = pressure at the upper level, $\dfrac{lb}{sq.\,ft.}$

R = radius of the earth = 20.89×10^6 ft.

m = molecular weight of the atmosphere.

T_m = the harmonic mean temperature in °R of the atmospheric layer of thickness $h_f - h_1$.

If the atmospheric layer from h_1 to h_f is divided into n equal parts or intervals the harmonic mean temperature T_m is defined by

$$\frac{1}{T_m} = \frac{1}{n} \left[\frac{1}{T_1} + \frac{1}{T_2} + \cdots\cdots + \frac{1}{T_n} \right], \tag{2}$$

(22) Humphreys, W.J.; Physics of the Air. McGraw Hill; pp.62-69, 1929.

Appendix A

where T_m is the average temperature in the nth interval. Thus to derive the vertical destribution of pressure, it is first necessary to have the temperature distribution curve, Fig. 3. Starting at some arbitrary level h_1 where the pressure p_1 is known, the pressure p_f at some higher level h_f is computed according to eq.(1). By dividing the atmosphere into a number of such layers the variation of the pressure with altitude is obtained. Since the NACA standard atmosphere was adopted hereto represent the lower part of the atmosphere up to 65,000 ft., this altitude served as the starting point of the pressure calculations.

Knowing the pressure and temperature, the density is computed from the equation of state.

$$\rho = \frac{pm}{g_o R_u T} , \qquad (3)$$

where $\rho =$ density, $\dfrac{\text{slugs}}{\text{ct.ft.}}$

$p =$ pressure, $\dfrac{\text{lbs}}{\text{sq.ft.}}$

$T =$ Temperature, $^\circ R$

$R_u =$ universal gas constant $= 1545$ $\dfrac{\text{lb-ft}}{\text{lb-mole}\ ^\circ F}$

$g_o =$ standard value for the acceleration of gravity $= 32.17$ $\dfrac{\text{ft}}{\text{sec}^2}$ (Constant)

$m =$ molecular weight of the atmosphere.

Using this system of units, the equations for computing the density is written.

$$\frac{pm}{49600T} : \qquad (4)$$

The adopted temperatures, from Fig. 3, and the corresponding
pressures and densities, computed as described above, are tabulated
in Table. 3.

TABLE 3
VALUES OF TEMPERATURE, PRESSURE, AND DENSITY OF THE ADOPTED UPPER ATMOSPHERE.

ALTITUDE (Thousands of feet)	KM.	Miles	Temperature, °R	Pressure, lbs / sq. ft.	Density, slugs / cu. ft.
0	0	0	518.4	2118	2.378×10^{-3}
35.3	10.76	6.64	392.4	490	7.27×10^{-4}
65	19.81	12.32	392.4	119.8	1.76×10^{-4}
100	30.48	18.95	500	26.01	3.06×10^{-5}
150	45.72	28.93	611	5.06	4.90×10^{-6}
200	60.96	37.90	628	1.17	1.11×10^{-6}
250	76.20	47.86	490	2.24×10^{-1}	2.67×10^{-7}
300	91.44	56.85	580	3.73×10^{-2}	3.75×10^{-8}
400	121.92	75.80	755	2.56×10^{-3}	1.97×10^{-9}
500	152.40	94.75	880	2.93×10^{-4}	1.78×10^{-10}
528	160.93	100.	906	1.81×10^{-4}	1.07×10^{-10}
600	182.88	113.70	984	5.28×10^{-5}	2.87×10^{-11}
700	213.36	132.65	1103	1.11×10^{-5}	4.38×10^{-12}
800	243.84	151.60	1264	2.89×10^{-6}	9.93×10^{-13}
900	274.32	170.55	1676	9.62×10^{-7}	2.49×10^{-13}
950	289.56	180.53	1788	6.16×10^{-7}	1.53×10^{-13}
1000	304.80	189.50	1798	4.00×10^{-7}	9.65×10^{-14}
1050	320.04	199.48	1700	2.58×10^{-7}	5.59×10^{-14}
1056	321.87	200.	1670	2.40×10^{-7}	6.24×10^{-14}
1100	335.28	208.45	1440	1.59×10^{-7}	4.79×10^{-14}
1200	365.76	227.40	1090	4.63×10^{-8}	1.84×10^{-14}
1300	396.74	246.35	950	1.06×10^{-8}	4.82×10^{-15}
1400	426.72	265.30	864	2.04×10^{-9}	1.03×10^{-15}
1500	457.20	284.25	810	3.55×10^{-10}	1.25×10^{-16}
1584	482.80	300.	774	1.58×10^{-10}	5.85×10^{-17}
1600	487.68	303.20	768	1.32×10^{-10}	4.93×10^{-17}
1700	518.16	322.15	760	4.75×10^{-11}	1.79×10^{-17}
1800	548.64	341.10	716	1.67×10^{-11}	6.66×10^{-18}
1900	579.12	360.05	700	5.75×10^{-12}	2.35×10^{-18}
2000	609.60	379.00	688	1.97×10^{-12}	8.20×10^{-19}
2100	640.08	397.95	674	6.69×10^{-13}	2.84×10^{-19}
2112	643.74	400.	671	5.80×10^{-13}	2.48×10^{-19}
2200	670.56	416.90	661	2.27×10^{-13}	9.80×10^{-20}
2300	701.04	435.85	656	7.66×10^{-14}	3.35×10^{-20}
2400	731.52	454.80	646	2.35×10^{-14}	1.04×10^{-20}
2500	762.00	473.75	640	7.93×10^{-15}	3.54×10^{-21}
2600	792.48	492.70	633	2.69×10^{-15}	1.22×10^{-21}
2640	804.61	500.	632	1.75×10^{-15}	7.90×10^{-22}
2700	822.96	511.65	630	9.21×10^{-16}	4.19×10^{-22}
2800	853.44	530.60	629	3.18×10^{-16}	1.44×10^{-22}
2900	883.92	549.55	627	1.11×10^{-16}	5.04×10^{-23}
3000	914.40	568.50	625	3.90×10^{-17}	1.79×10^{-23}
3100	944.88	587.45	624	1.39×10^{-17}	6.36×10^{-24}
3168	965.60	600.	623	7.00×10^{-18}	3.21×10^{-24}
3200	975.36	606.40	620	4.98×10^{-18}	2.30×10^{-24}
3300	1005.84	625.35	617	1.81×10^{-18}	8.37×10^{-25}

Table 3
Appendix A

TABLE 4
ALTITUDE CONVERSION

KM.	ALTITUDE THOUSANDS OF FEET	MILES
0	0	0
20	65.6	12.4
40	131.2	24.9
60	196.8	37.3
80	262.5	49.7
100	328.1	62.1
120	393.7	74.6
140	459.3	87.0
160	524.9	99.4
180	590.6	111.8
200	656.2	124.3
220	721.8	136.7
240	787.4	149.1
260	853.0	161.6
280	918.6	174.0
300	984.2	186.4
320	1050.	198.8
340	1115.	211.3
360	1181.	223.7
380	1247.	236.1
400	1312.	248.5
420	1378	261.0
440	1444.	273.4
460	1509	285.8
480	1575	298.3
500	1640	310.7

KM.	ALTITUDE THOUSANDS OF FEET	MILES
520	1706.	323.1
540	1772.	335.5
560	1837.	348.0
580	1903.	360.4
600	1968.	372.8
620	2034.	385.2
640	2100.	397.7
660	2165.	410.1
680	2231.	422.5
700	2297.	435.0
720	2362.	447.4
740	2428.	459.8
760	2493.	472.2
780	2559.	484.7
800	2625.	497.1
820	2690.	509.5
840	2756.	298.2
860	2822.	534.4
880	2887.	546.8
900	2953	559.2
920	3018.	571.7
940	3084.	584.1
960	3150.	596.5
980	3215.	608.9
1000	3281.	621.4

MILES	ALTITUDE THOUSANDS OF FEET
0	0
25	132
50	264
75	396
100	528
125	660
160	792
175	924
200	1056
225	1188
250	1320
275	1452
300	1584
325	1716
350	1848
375	1980
400	2112
425	2244
450	2376
475	2508
500	2640
525	2772
550	2904
575	3036
600	3168

1 KM. = 3280.8 FT.

Table 4
Appendix A

TABLE 5
TEMPERATURE CONVERSION

°K	°C	°F	°R
0			
20	-253	-423.4	36.6
40	-233	-387.4	72.6
60	-213	-351.4	108.6
80	-193	-315.4	144.6
100	-173	-279.4	180.6
120	-153	-243.4	216.6
140	-133	-207.4	252.6
160	-113	-171.4	288.6
180	-93	-135.4	324.6
200	-73	-99.4	360.6
220	-53	-63.4	396.6
240	-33	-27.4	432.6
260	-13	8.6	468.6
280	7	44.6	504.6
300	27	80.6	540.6
320	47	116.6	576.6
340	67	152.6	612.6
360	87	188.6	648.6
380	107	224.6	684.6
400	127	260.6	720.6
420	147	296.6	756.6
440	167	332.6	792.6
460	187	368.6	828.6
480	207	404.6	864.6

°K	°C	°F	°R
500	227	440.6	900.6
520	247	476.6	936.6
540	267	512.6	972.6
560	287	548.6	1008.6
580	307	584.6	1044.6
600	327	620.6	1080.6
620	347	656.6	1116.6
640	367	692.6	1152.6
660	387	728.6	1188.6
680	407	764.6	1224.6
700	427	800.6	1260.6
720	447	836.6	1296.6
740	467	872.6	1332.6
760	487	908.6	1368.6
780	507	944.6	1404.6
800	527	980.6	1440.6
820	547	1016.6	1476.6
840	567	1052.6	1512.6
860	587	1088.6	1548.6
880	607	1124.6	1584.6
900	627	1160.6	1620.6
920	647	1196.6	1656.6
940	667	1232.6	1692.6
960	687	1268.6	1728.6
980	707	1304.6	1764.6
1000	727	1340.6	1800.6

$$°K = °C + 273$$
$$°R = °F + 460$$

Table 5
Appendix A

Appendix B

B.   THE DETERMINATION OF THE DRAG COEFFICIENT.

In order to simplify the estimation of the drag coefficient for the purposes of this report, it was assumed that the vehicle was effectively conical in shape. Since drag, in almost all cases, is a second order effect, such an assumption is not out of order. The half cone angle, θ, was taken to be 0.3 radians. Other values of θ however, were eventually chosen for the more final designs.

At subsonic speeds, the total of pressure, friction, and base drag coefficients was taken to be 0.3, where the definition of C_D is

$$C_D = \frac{\text{Drag}}{(1/2)\rho V^2 A}$$

V is the velocity.

ρ is the mass density of air.

A is the frontal area.

This value was held constant for $0 < M < 0.8$.

At low supersonic speeds, the well known work of Taylor and Maccoll is available, and gives values which are shown on the curve on fig. 1 . Kinetic theory, under the assumption of inelastic impacts destroying the normal component of momentum, gives $2\theta^2$ for the hypersonic pressure drag coefficient. The supersonic base pressure coefficient, $(2/3)\frac{1}{\gamma M^2}$, was reduced to $\frac{.3}{M^2}$ because the base area of the rocket jet did not contribute drag. Skin friction, when based on frontal area, was sufficiently small to ignore (0.020) at high Reynolds numbers.

In the transonic region, a total drag coefficient of .45 was selected. This value is also shown on the accompanying graph, fig. 1.

| 25 B9
.EV. 7.42Y

Analysis _DRAG OF CONE_

Prepared by _HAROLD LUSKIN_

Date _5 - 6 - 46_

DOUGLAS AIRCRAFT COMPANY, INC.

SANTA MONICA

Plant

Page _2B_

Model _#1033_

Report No. _SM 1182_

DRAG COEFFICIENTS FOR A CONE WITH HALF CONE ANGLE OF 0.3 RADIANS

TOTAL DRAG AT EXTREMELY LOW REYNOLDS NUMBERS

TOTAL DRAG AT HIGH REYNOLDS NUMBERS

KINETIC THEORY

TAYLOR - MACCOLL

PRESSURE DRAG

BASE DRAG

MACH NUMBER

C_D

Appendix B

The values discussed above were used in the calculations for the
ascent to the orbital height. The importance of the drag in those calculations gradually decreased, until at about 150,000 ft. above sea level,
it was negligible compared to the thrust.

In cases where long periods of time are involved, especially where
also the Reynolds numbers are low, it is necessary to consider carefully
drags which would otherwise seem to be negligible. The detailed analyses
of Sänger and others in Germany have given results for high altitude, high
M conditions similar to those obtained by considering the destruction of
all momentum in a cylinder whose cross-section is the same as that of the
vehicle, and which approaches the vehicle at the vehicles speed. Under
such assumptions, it is found that $C_D = 2.0$.

As discussed elsewhere in this report, there are four regimes of
flow which can conveniently be characterized by the ratio of mean free
path to boundary layer thickness, or by the ratio of Mach number to square
root of Reynolds number. The plot on fig. 2 shows these regimes and where
in these various realms, the space vehicle flies. It will be noted that
on fig. 1 no drag data are given for the slip or unknown regions. In
these regions much research must be done.

Analysis .. DRAG

Prepared by CLAEYS

Date   4 - 4 - 46

DOUGLAS AIRCRAFT COMPANY, INC.

Page 4B

Model #1033

Report No. SM. 11827

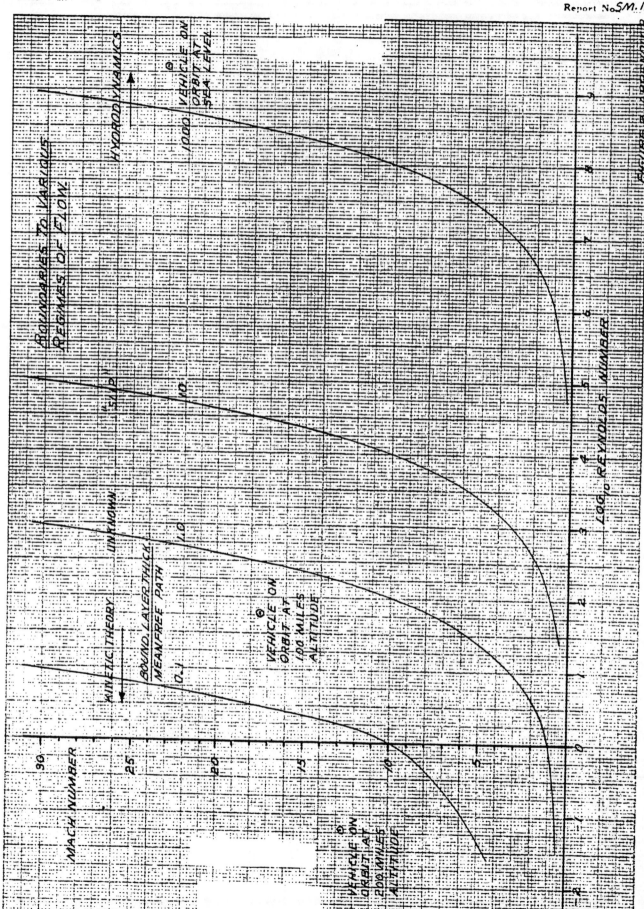

BOUNDARIES TO VARIOUS REGIMES OF FLOW

HYDRODYNAMICS

1000 VEHICLE ON ORBIT AT SEA LEVEL

"SLIP"

KINETIC THEORY

UNKNOWN

BOUND. LAYER THICK. MEAN FREE PATH

VEHICLE ON ORBIT AT 100 MILES ALTITUDE

VEHICLE ON ORBIT AT 200 MILES ALTITUDE

MACH NUMBER

LOG₁₀ REYNOLDS NUMBER

FIGURE 8 DRAG

Appendix C

C. LAGRANGIAN EQUATIONS

Development of the Equations of Motion of a Body Moving at Great Speeds Near the Surface of the Earth in the Plane of the Equator. In this appendix, the equations of motion of a body will be developed in a form suitable for use in calculating the trajectory followed by the body as it is accelerated to the proper speed and direction for orbital motion. The analysis is confined to motion in the plane of the earth's equator because this is the only case considered in the calculations of the main text.

We shall take the following as variables characterizing the motion of the body: r, the radial distance to the body from the center of the earth; φ, the longitudinal angle of the body; and t, the time. We shall call m, the mass of the body; g, the acceleration of gravity; k, the gravitational constant; M, the earth's mass; R, the radius of the earth; and Ω, the angular velocity of the earth.

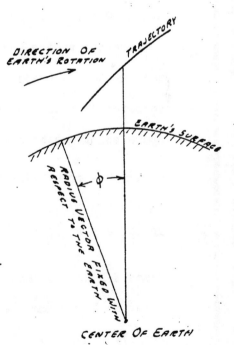

The kinetic energy of the body is

$$T = \frac{m}{2}\left[\dot{r}^2 + r^2(\dot{\varphi} + \Omega)^2\right]$$

and the potential energy is

$$U = -\frac{kmM}{r}$$

Appendix C

If we form the La Grangian function $L = T - U$, then

$$\frac{d}{dt}\frac{\partial L}{\partial \dot{r}} - \frac{\partial L}{\partial r} = m\ddot{r} - mr(\dot{\varphi}+\Omega)^2 + \frac{kmM}{r^2} = F_r$$

and

$$\frac{d}{dt}\frac{\partial L}{\partial \dot{\varphi}} - \frac{\partial L}{\partial \varphi} = 2mr\dot{r}(\dot{\varphi}+\Omega) + mr^2\ddot{\varphi} = rF_\varphi$$

where F_r and F_φ are the radial and tangential components of all the externally applied forces.

We now rotate our coordinate system so that the vector resolutions are parallel and perpendicular to the trajectory. We designate by **v** the velocity of the body measured from a frame of reference-fixed in the earth and by Θ the complement of the angle between the radius vector and the tangent to the trajectory measured in coordinates fixed in the earth. The above equations become

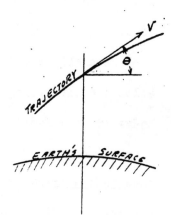

$$m\ddot{r}\sin\Theta - mr(\dot{\varphi}+\Omega)^2\sin\Theta + 2m\dot{r}(\dot{\varphi}+\Omega)\cos\Theta + mr\ddot{\varphi}\cos\Theta + \frac{kmM\sin\Theta}{r^2} = T-D,$$

$$mr\cos\Theta - mr(\dot{\varphi}+\Omega)^2\cos\Theta - 2mr(\dot{\varphi}+\Omega)\sin\Theta - mr\ddot{\varphi}\sin\Theta + \frac{kmM\cos\Theta}{r^2} = L,$$

where T-D is the thrust minus the drag, along the trajectory and L is the lift normal to the trajectory (positive when in the direction $+ \pi/2$).

Appendix C

We now eliminate derivatives of r and φ by noting the following:

$$\dot{r} = v \sin \theta \qquad\qquad \dot{\varphi} = \frac{v \cos \theta}{r}$$

$$\ddot{r} = \dot{v} \sin \theta + v \cos \theta \, \dot{\theta} \qquad \ddot{\varphi} = \frac{\dot{v} \cos \theta}{r} - \frac{v \sin \theta \, \dot{\theta}}{r} - \frac{v^2 \sin \theta \cos \theta}{r^2}$$

These give, upon substitution in the equations of motion

$$m \dot{v} - m r \Omega^2 \sin \theta + \frac{kmM \sin \theta}{r^2} = T - D$$

$$m v \dot{\theta} - \frac{m v^2 \cos \theta}{r} - 2 m v \dot{\varphi} - m r \dot{\varphi}^2 \cos \theta + \frac{kmM \cos \theta}{r^2} = L$$

We can readily evaluate the gravitational constant because we know that when the body is standing still on the earth's surface, an external force mg is required to keep it in equilibrium. Putting $v = \dot{v} = \dot{\theta} = \theta = T - D = 0$ and $L = mg$ and $r = R$, we have

$$- m R \Omega^2 + \frac{k m M}{R^2} = m g \quad \text{or} \quad k M = g R^2 + R^3 \Omega^2,$$

Substituting this value of kM back into the equations of motion, we have

$$m \dot{v} - m r \Omega^2 \sin \theta + \frac{m(g R^2 + R^3 \Omega^2)\sin \theta}{r^2} = T - D$$

$$m v \dot{\theta} - \frac{m}{r} \cos \theta \, (v + r \Omega)^2 - 2m v \Omega (1 - \cos \theta) + \frac{m(g R^2 + R^3 \Omega^2)\cos \theta}{r^2} = L$$

It is of interest to investigate the significance of the various terms in these equations. In the first equation, which represents an equilibrium of forces in the direction of motion, $m \dot{v}$, T and D are the factors entering the familiar mass x acceleration = force. The term, $m r \Omega^2 \sin \theta$ is the component in the direction of motion of the centrifugal force caused by the earth's rotation. The term $\frac{m(g R^2 + R^3 \Omega^2)\sin \theta}{r^2}$

Appendix C

is the corresponding component of the earth's attraction when account is taken of the fact that this attraction will impart an acceleration of 1g in the presence of the earth's rotation. In the second equation, which represents equilibrium of forces perpendicular to the direction of motion, in the plane of the equator, the term $m v \dot{\theta}$ is the centrifugal force as seen from local earth coordinates. The term $\frac{m \cos \theta}{r} (v + r\Omega)^2$ is the component of centrifugal force of the total rotation around the earth. The term, $2 m v (1 - \cos \theta)$ can be interpreted as the apparent force that causes a body, ejected outward from the earth, to be left behind as the earth rotates under it. The term $\frac{m(g R^2 + R^3 \Omega^2)}{r^2} \cos \theta$ is, of course, the component of the earth's attraction to normal to the direction of motion.

Using the above equations, we shall investigate the simple case of the free motion of the vehicle in a circular orbit at the earth's surface neglecting air resistance. For this case we put $L = \theta = \dot{\theta} = 0$; $R = r$ and the orbital velocity is determined by the relations

$$\frac{v^2}{R} + 2 v\Omega + R \Omega^2 - g - R\Omega^2 = 0,$$

$$v = R\Omega \overset{+}{}{-} \sqrt{(R\Omega)^2 + g R} .$$

Using an equatorial radius of 3,963 miles and a value of $g = 32.086$, $v = - 24,319$ ft./sec. and 27,369 ft./sec. depending on whether the vehicle is moving with or against the earth's rotation.

It is seen that these values differ only slightly from the values of -24,285 ft./sec. and 27,335 ft./sec. given in Chapter III.

Appendix C

The values of Chapter III were computed from a simplified formula $v = R\Omega \pm \sqrt{gR}$ which neglects the effect of the earth's rotation on the apparent gravitational attraction of a stationary object. This difference in attraction is small, amounting to only about .6%.

Returning to the equation of motion to be used for the trajectory calculations, we shall put the altitude, $h = r - R$. For a trajectory 100 miles high, $\frac{h}{R} = 2\text{-}1/2\%$, a quantity whose square can be neglected compared to unity. Using this approximation, the equations take the form

$$\frac{dv}{dt} = 3H\,\Omega^2 \sin\theta - g\left(1 - 2\frac{h}{R}\right)\,\sin\theta + \frac{T - D}{m},$$

$$\frac{d\theta}{dt} = \frac{v}{R}\left(1 - \frac{h}{R}\right)\cos\theta + 2\,\Omega + 3\,\frac{h\Omega^2}{v}\cos\theta - \frac{g\left(1 - \frac{2h}{R}\right)}{v}\cos\theta + \frac{L}{m\,v}$$

If the terms in the above equation are examined for order of magnitude it is seen that $3h\,\Omega^2$ is always small compared to g. Consequently, we can maintain an accuracy of better than 1% using the following

$$\frac{dv}{dt} = -g\left(1 - \frac{2h}{R}\right)\,\sin\theta + \frac{T - D}{m},$$

$$\frac{d\theta}{dt} = \frac{v}{R}\left(1 - \frac{h}{R}\right)\cos\theta + 2\,\Omega - \frac{g\left(1 - \frac{2h}{R}\right)}{v}\cos\theta + \frac{L}{m\,v}$$

In the step by step calculations of trajectories, a preliminary calculation was usually made neglecting $\frac{h}{R}$ compared to unity and later the results were corrected for these small terms.

Appendix   D

D. SAMPLE OF THE DETAILED TRAJECTORY CALCULATION

A = Maximum cross-sectional area of vehicle,

D = Drag,

g = Acceleration of gravity at earth's surface,

h = Altitude,

I = Specific impulse,

r = Distance from earth's center to vehicle,

R = Radius of earth,

t = Time

t_B = Burning time,

V = Velocity of vehicle,

V_E = Circumferential velocity of earth

W = Instantaneous mass of vehicle,

x = Distance projected on earth's surface,

γ = $\dfrac{\text{Original fuel weight per stage,}}{\text{Original total weight per stage}}$

α = Angle thrust makes with flight path (tilt),

θ = Angle of inclination of flight path to earth's horizontal.

Subscripts

l = (lower) first three burning periods,

μ = (upper) fourth burning period,

c = Coasting,

0 = Initial condition,

1 = Beginning of an interval,

2 = End of an interval.

Appendix D

It is our purpose here to discuss the methods of calculating the trajectory. Neglecting $\frac{h}{R}$, the equations of motion are (see appendix C):

(1) $\quad \dfrac{dV}{dt} = \dfrac{g\,I\,\cos\alpha}{t_B\left(1-\gamma\frac{t}{t_B}\right)} - g\sin\theta - g\,\dfrac{D}{W}$

(2) $\quad \dfrac{d\theta}{dt} = -g\cos\theta + \dfrac{V^2}{R}\cos\theta + \dfrac{2VV_E}{R} - \dfrac{g\,I}{t_B\left(1-\gamma\frac{t}{t_B}\right)}\sin\theta,$

(3) $\quad \dfrac{dh}{dt} = V\sin\theta,$

(4) $\quad \dfrac{dx}{dt} = V\cos\theta.$

When the term $\dfrac{g\,I\,\gamma}{t_B\left(1-\gamma\frac{t}{t_B}\right)}$ and $g\,\dfrac{D}{W}$ due to thrust and drag are

absent, $\frac{h}{R}$ cannot be neglected. However, normally it can be

neglected because $\frac{h}{R}$ occurs only in terms which are small compared to the

rocket thrust terms.

We shall first review the general method of calculation, using as an example the four-stage alcohol-oxygen trajectory. Methods for the other cases presented are similar. The vehicle travels vertically for half the time of the first burning period after which tilt is applied. The tilt remains constant until the end of the third burning period at which time coasting begins. In the last burning period a new angle of tilt is held constant. Because a knowledge of altitude is required in the calculations for the first and second periods and coasting, these computations were made as a group beginning at sea level and ending after coasting. Because

Appendix D

the final velocity conditions are to a first approximation independent of altitude, we calculate the fourth period backward. From the first set of calculations, we obtain $\Delta\theta_\ell, \Delta V_\ell, \Delta h_\ell$ at the beginning of coasting for various values of γ and α_ℓ (plots shown in Figure D1, D2, D3). From the second set of calculations, plots (Fig. D4, D5, D6) of $\Delta\theta_u, \Delta V_u, \Delta h_u$ versus γ and α_u are made. The coasting trajectories were computed from the equations for elliptic orbits which are discussed later.

To determine an actual trajectory, these plots and the coasting calculations are used to solve simultaneously the following equations:

$$\Delta V_\ell + \Delta V_c + \Delta V_u = \text{Orbital speed (depends on altitude)}$$

$$\Delta\theta_\ell + \Delta\theta_c + \Delta\theta_u = 90^\circ$$

$$\Delta h_\ell + \Delta h_c + \Delta h_u = \text{Desired altitude}$$

With four independent variables (γ, α_ℓ, α_u and Δh_c) and three restraining conditions, (orbital velocity, direction and altitude) we seek graphically the optimum trajectory for the γ of the proposed designs.

The actual step by step calculations in the burning periods differed somewhat between the burning periods. In the first burning period equation (2) had very little effect on equation (1) so that its effect on V as a function of t could be handled as a small perturbation of the results of (1) after its completion. In spite of this simplification, the calculations in the first burning period were the most difficult since the drag was large and the variation of I was considerable. If (1) is integrated for an interval in which I and $\frac{D}{W}$ could be considered constant, we have

Appendix D

$$V_2 - V_1 = -g\,\overline{I}\cos\alpha\,\log\frac{1-\gamma\frac{t_2}{t_B}}{1-\gamma\frac{t_1}{t_B}} - g(t_2-t_1) - g\frac{\overline{D}}{W}(t_2-t_1)$$

A second integration yields

$$h_2 - h_1 = (V_1 + g\overline{I})(t_2-t_1) + g\overline{I}\frac{t_B}{\gamma}\left(1-\gamma\frac{t_2}{t_B}\right)\log\left(\frac{1-\gamma\frac{t_2}{t_B}}{1-\gamma\frac{t_1}{t_B}}\right) - g\left(1+\frac{\overline{D}}{W}\right)\frac{(t_2-t_1)^2}{2}$$

The second equation is used to determine the altitude necessary for a knowledge of I and $\frac{B}{W} = \frac{C_D q\,A}{W_o(1-\gamma\frac{t}{t_o})}$ where I and the density in q depend only on altitude. The fact that (1) can be thus integrated enables us to take much larger steps than a complete iteration process would require.

Having established V as a function of t, we can use it in integrating (2) in a similar fashion:

$$\Delta\theta = \left(-\frac{g\,\overline{\cos\theta}}{\overline{V}} + \frac{\overline{V}\,\overline{\cos\theta}}{R} + \frac{2V_E}{R}\right)\Delta t + \frac{g\,\overline{I}}{\overline{V}}\log\frac{1-\gamma\frac{t_2}{t_B}}{1-\gamma\frac{t_1}{t_B}}$$

from this, we obtain θ as a function of t to correct the original velocity calculations.

In the second burning period I is constant and $\frac{D}{W}$ is negligible over most of the period. However, the variation in θ is now appreciable so that equations (1) and (2) must be iterated simultaneously.

The following remarks apply for all burning periods except the first. The partly integrated equations for a small interval are:

$$(1)'\quad \Delta V = -g\,\overline{I}\cos\alpha\,\log\left(\frac{1-\gamma\frac{t_2}{t_B}}{1-\gamma\frac{t_1}{t_B}}\right) - g\,\overline{\sin\theta}\,\Delta t$$

Appendix   D

$$(2)' \quad \Delta\theta = \left(-g\,\frac{\overline{\cos\theta}}{\overline{V}} + \frac{V}{R}\,\overline{\cos\theta} + 2\,\frac{V_E}{R} \right)\Delta t + g\,\overline{I}\,\sin\alpha\,\log\frac{1-\gamma\frac{t_2}{t_B}}{1-\gamma\frac{t_1}{t_B}}$$

Starting with (1)' we calculate ΔV to use in (2)' for $\overline{V}$ from which we get $\Delta\theta$ to use in (1)' for $\overline{\sin\theta}$ to get a better value of V. Altitudes need not be calculated until V and θ as functions of t are established and then they and the r's may be obtained by planimeter from equations (3) and (4).   Sample calculations from each burning period are presented.

For coasting (rocket thrust absent) $\frac{h}{R}$ is not negligible so (1) and (2) become

$$\frac{dV}{dt} = -g\left(\frac{R}{r}\right)^2\sin\theta,$$

$$V\,\frac{d\theta}{dt} = -g\left(\frac{R}{r}\right)^2\cos\theta + \frac{V^2}{r}\cos\theta + 2\,\frac{VV_E}{R}$$

If the $\frac{V_E}{R}$ term is neglected these equations integrate into

$$V_2{}^2 - V_1{}^2 = -2gR\left(\frac{R}{r_1} - \frac{R}{r_2}\right)$$

and $\dfrac{\cos\theta_2}{\cos\theta_1} = \dfrac{r_1}{r_2}\dfrac{V_1}{V_2}$, which are the equations of motion of a body in an

elliptic orbit. When account is taken of the effect of the earth's motion the second equation is modified for sufficient accuracy into

$$\frac{\cos\theta_2}{\cos\theta_1} = \frac{r_1}{r_2}\,\frac{V_1}{V_2}\,e^{-\frac{2V_E}{V\cos\theta}\frac{\Delta r}{R}}$$

These equations are in the form used in the calculations.

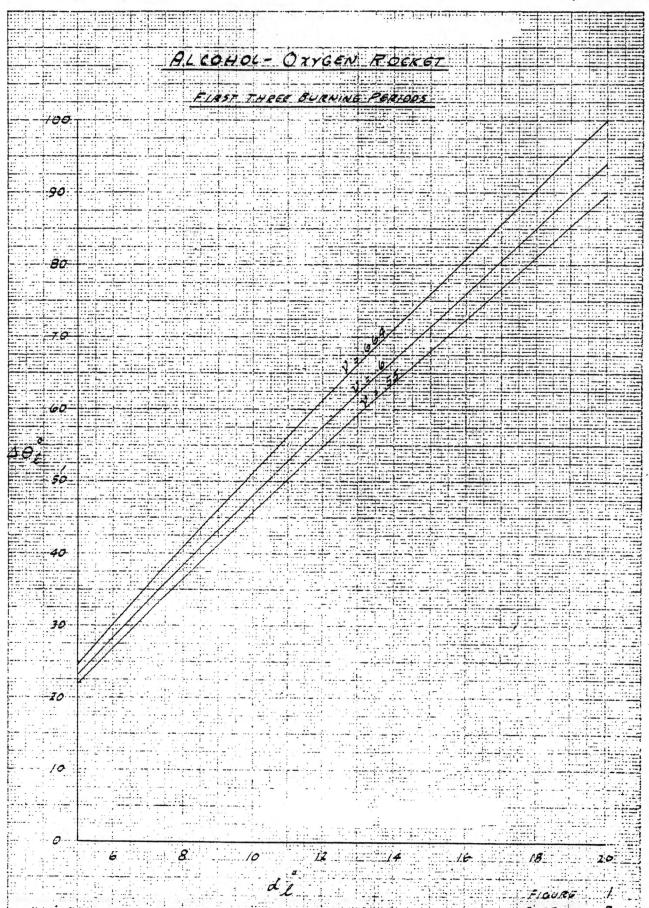

ALCOHOL - OXYGEN ROCKET

FIRST THREE BURNING PERIODS

FIGURE   4

APPENDIX D

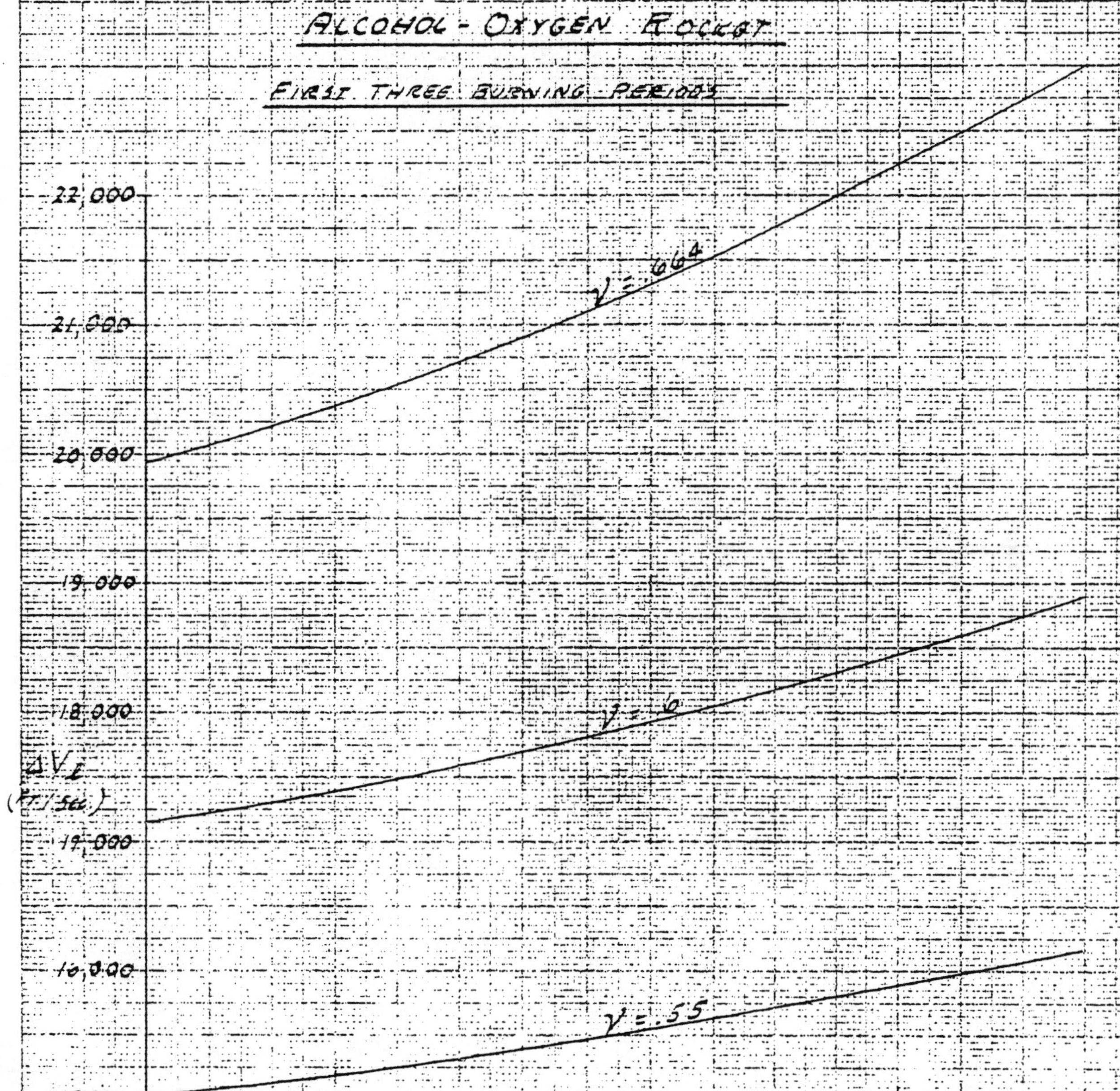

Analysis PRELIMINARY DESIGN OF SATELLITE VEHICLE

Prepared by W. H. Wampler   DOUGLAS AIRCRAFT COMPANY, INC.

Date   May 2, 1946

Page 7D

Model #1033

Report No. SM11827

ALCOHOL - OXYGEN ROCKET

FIRST THREE BURNING PERIODS

$\gamma = .66$

$\gamma = .6$

$\gamma = .55$

ΔV_i (ft./sec.)

d_i^o

FIGURE 2

APPENDIX II

ORM 25 BP

Analysis PRELIMINARY DESIGN OF SATELLITE VEHICLE

Prepared by W. H. Wampler   DOUGLAS AIRCRAFT COMPANY, INC.

Date   May 2, 1946

Page 8D

Model #1022

Report No. SM11827

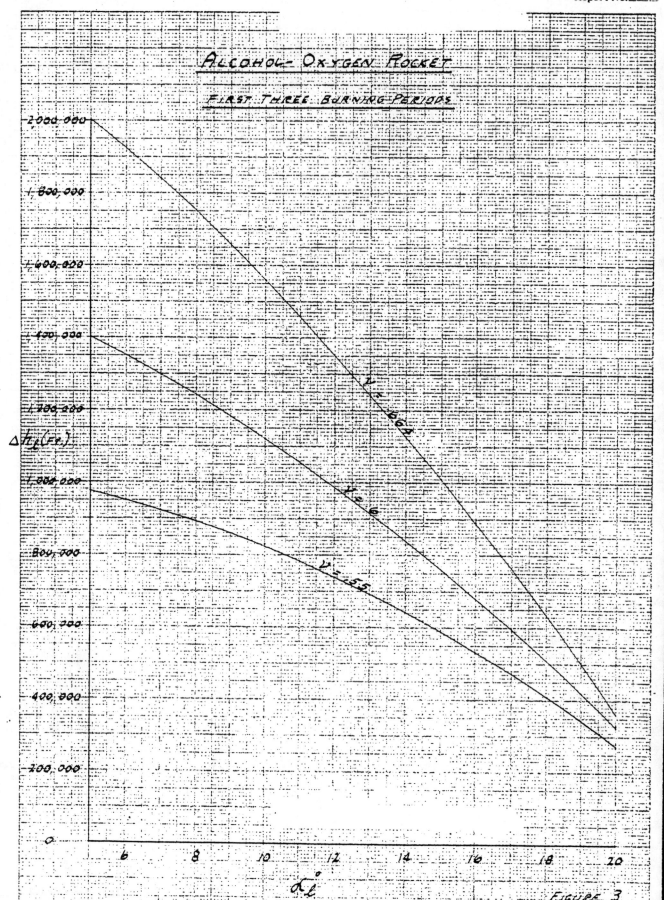

ALCOHOL- OXYGEN ROCKET

FIRST THREE BURNING PERIODS

FIGURE 3

Analysis PRELIMINARY DESIGN OF SATELLITE VEHICLE
Prepared by. h. E. leggler
DOUGLAS AIRCRAFT COMPANY, INC.
Date May 2, 1946
Page 9D
Model #1035
Report No. SM11827

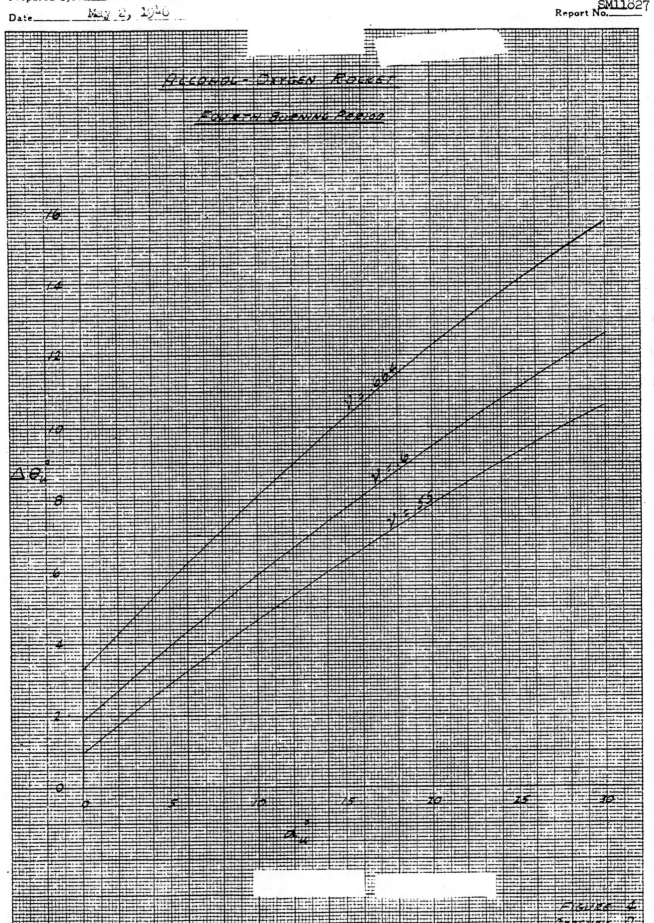

FIGURE 4
APPENDIX I

FORM 25 BP

Analysis PRELIMINARY DESIGN OF SATELLITE VEHICLE

Prepared by W. H. Wampler   DOUGLAS AIRCRAFT COMPANY, INC.

Date May 2, 1946

Page 10D

Model #1033

Report No. SM11627

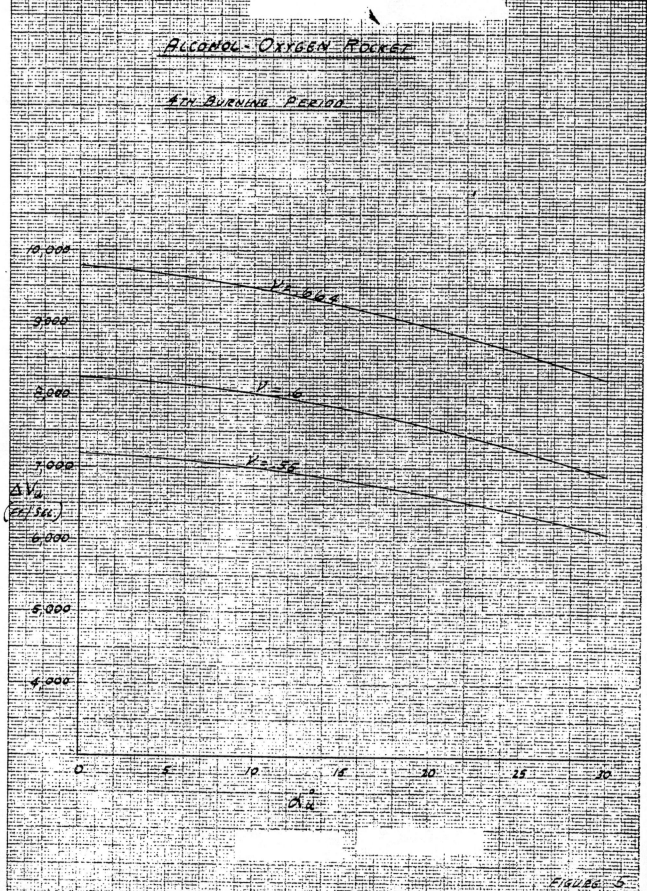

FIGURE 5

APPENDIX D

Analysis PRELIMINARY DESIGN OF SATELLITE VEHICLE

Prepared by W. H. Wampler   DOUGLAS AIRCRAFT COMPANY, INC.

Date   May 2, 1946

Page 11D

Model #1033

Report No. SM11827

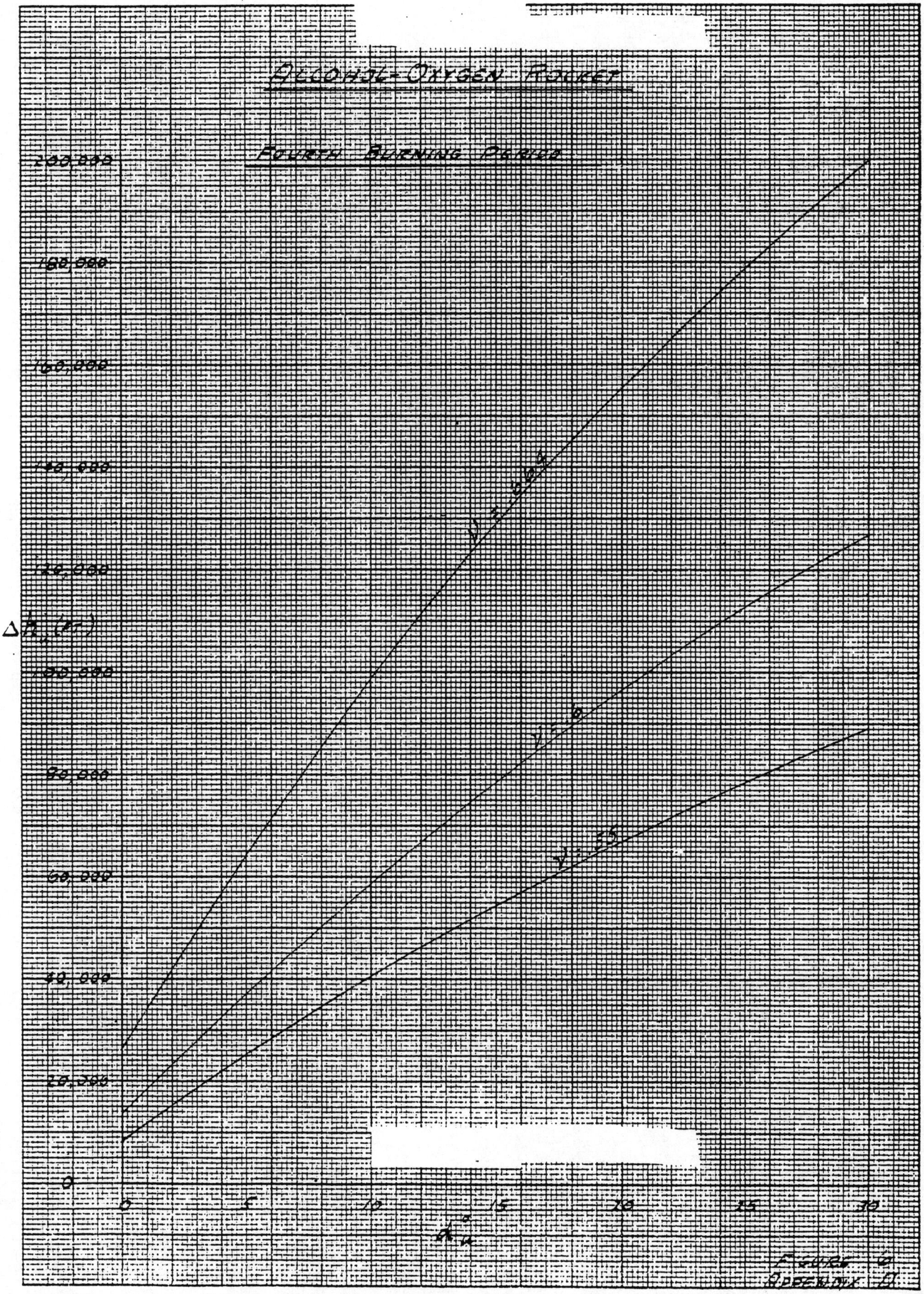

FORM 25 GP

Figure 9
Appendix III

Analysis PRELIMINARY DESIGN OF SATELLITE VEHICLE
Prepared by W. H. Wampler   DOUGLAS AIRCRAFT COMPANY, INC.
Date May 2, 1946

Page 12D
Model #1033
Report No. SM11827

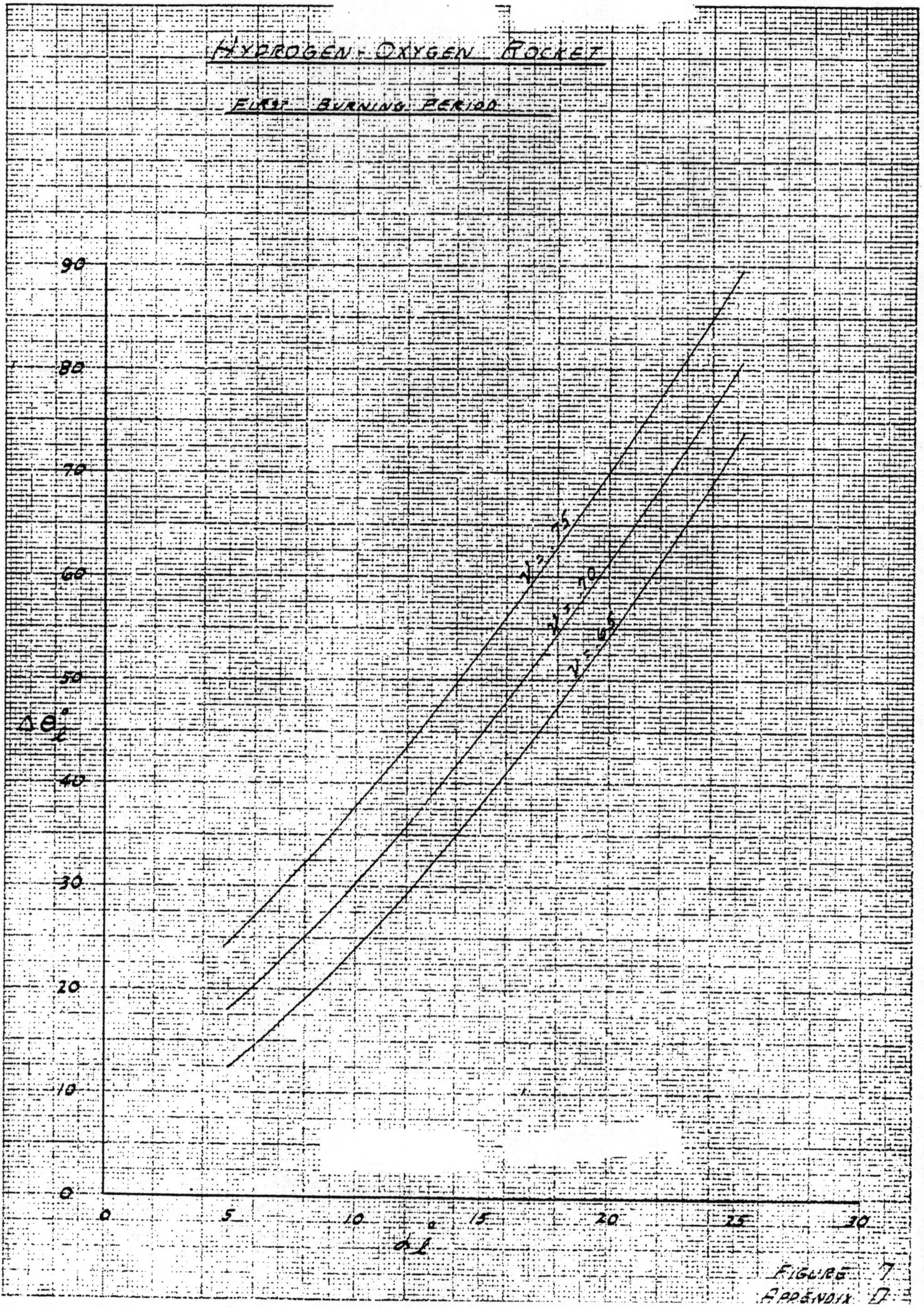

HYDROGEN-OXYGEN ROCKET

FIRST BURNING PERIOD

FIGURE 7
APPENDIX D

FORM 25 BP

Analysis _PRELIMINARY DESIGN OF SATELLITE VEHICLE_   Page _13D_
Prepared by _W. H. Wampler_   DOUGLAS AIRCRAFT COMPANY, INC.   Model _#1033_
Date _May 2, 1946_   Report No. _SM11827_

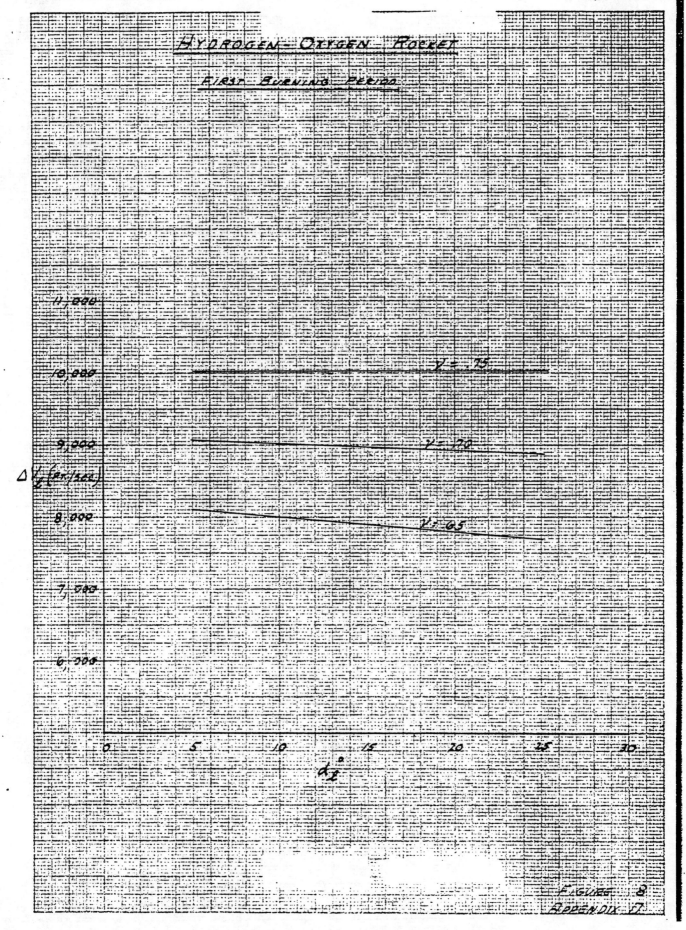

FIGURE 8
APPENDIX VI

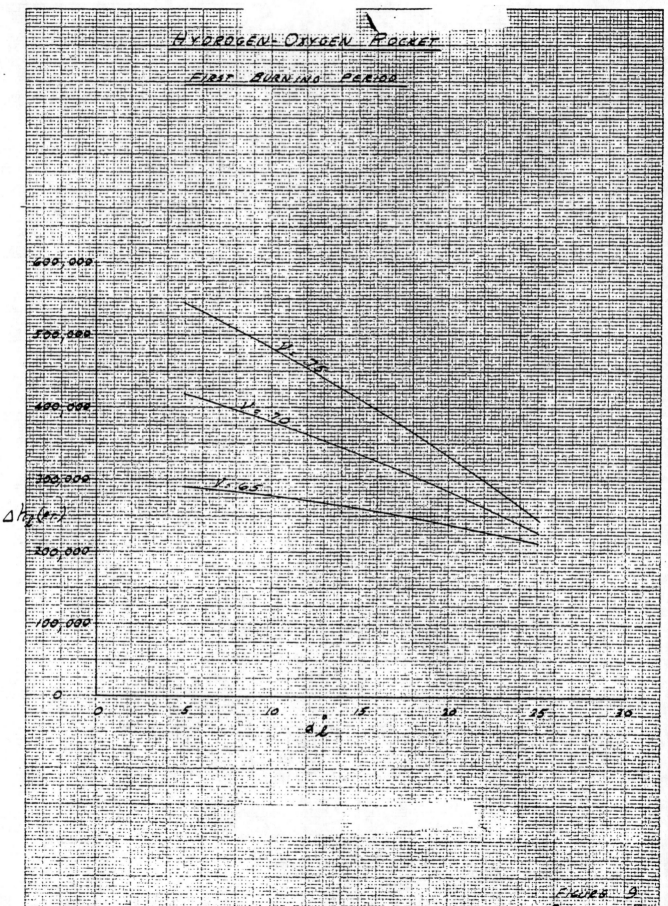

HYDROGEN-OXYGEN ROCKET

FIRST BURNING PERIOD

FIGURE 9

APPENDIX D

Analysis PRELIMINARY DESIGN OF SATELLITE VEHICLE

Prepared by W. H. Wampler   DOUGLAS AIRCRAFT COMPANY, INC.

Date May 2, 1946

Page 15D

Model #1033

Report No. SM11827

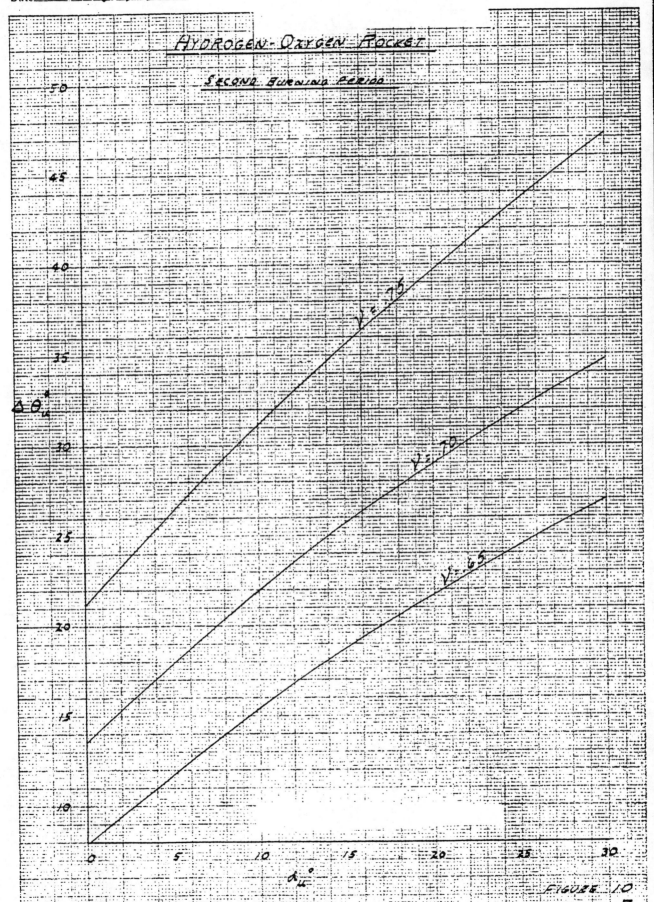

FIGURE 10

Analysis PRELIMINARY DESIGN OF SATELLITE VEHICLE
Prepared by W. H. Wampler
Date May 2, 1946

DOUGLAS AIRCRAFT COMPANY, INC.

Page 16D
Model #1033
Report No. SM11827

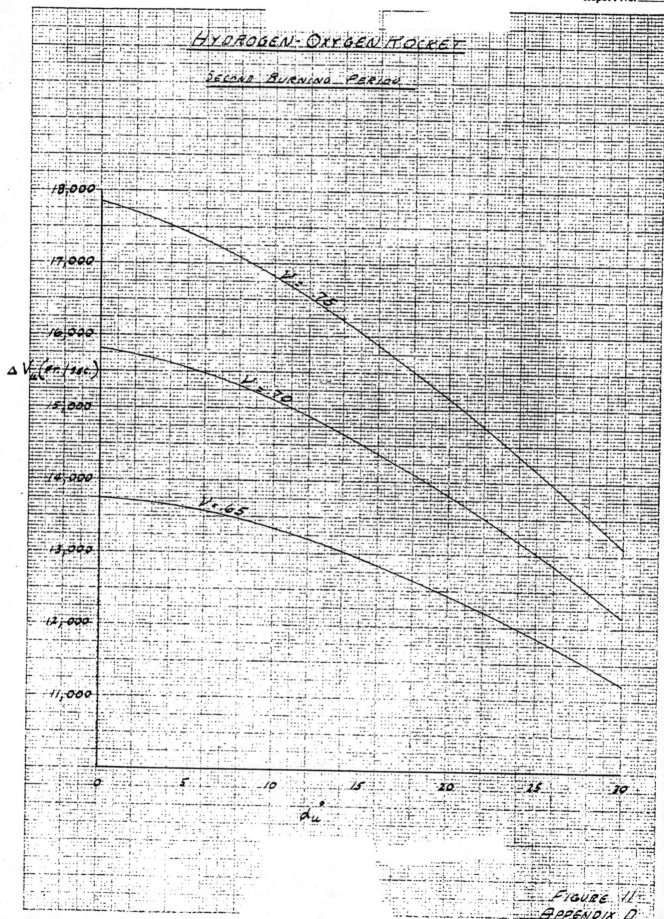

HYDROGEN-OXYGEN ROCKET

SECOND BURNING PERIOD

FIGURE #1
APPENDIX D

FORM 25 BP

Analysis PRELIMINARY DESIGN OF SATELLITE VEHICLE

Prepared by W. H. Wampler

DOUGLAS AIRCRAFT COMPANY, INC.

Date May 2, 1946

Page 17D

Model #1033

Report No. SM-11827

HYDROGEN - OXYGEN ROCKET

SECOND BURNING PERIOD

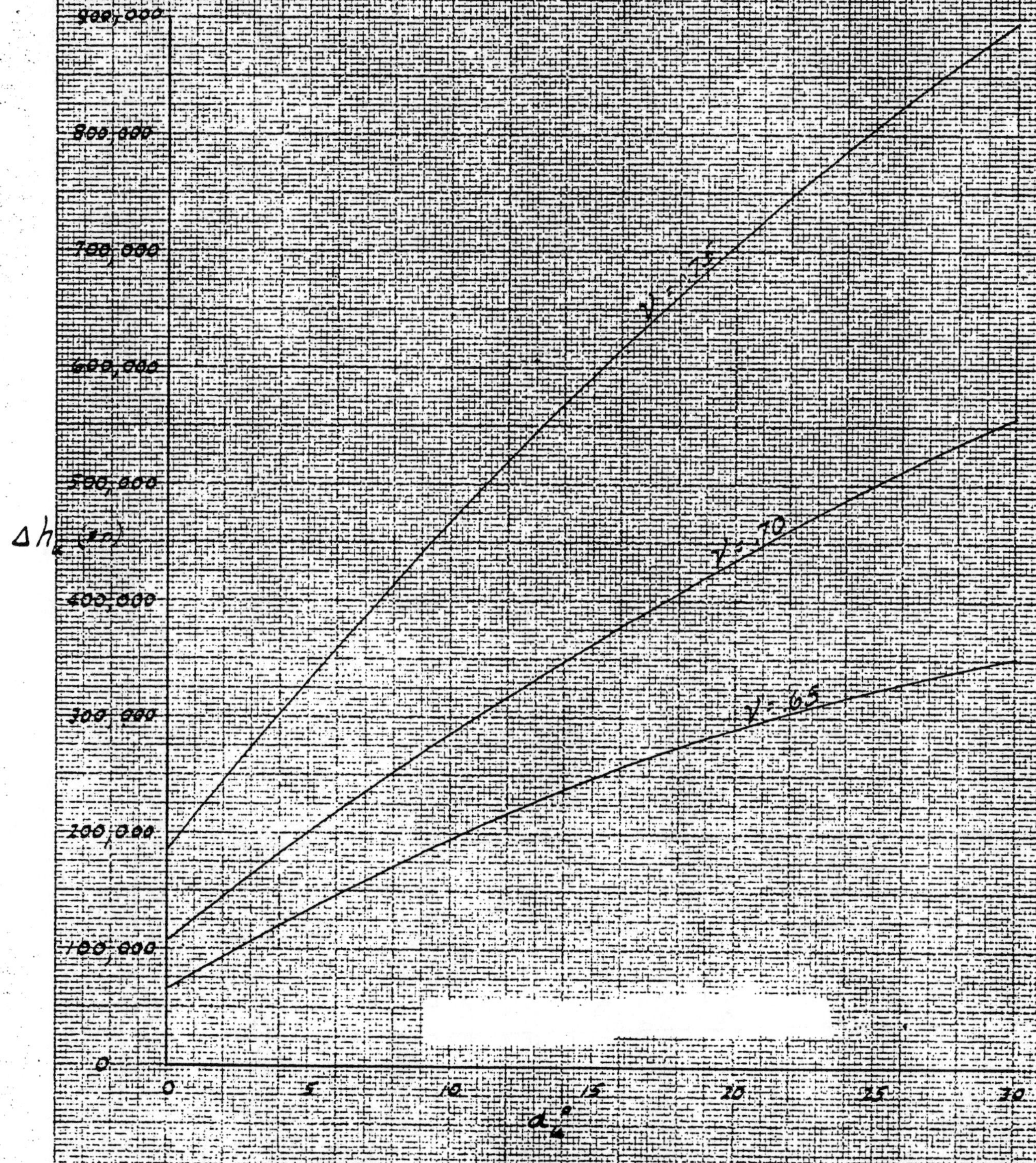

FIGURE 12
APPENDIX A

WORK SHEET
FORM 30-103-1 (REV. 1-43)

1st BURNING PERIOD

$\gamma = .55 \qquad \alpha = 15°$

t (sec)	δt	$1-\nu\frac{t}{t_k}$	$\log(1-\nu\frac{t}{t_k})$	$\log(1-\nu^2\frac{t}{t_k})$	I	Ī		$-g\delta t$		δV	CORRECTION DUE TO θ	V	$\delta t(\nu_k+g\tau)$	$\frac{g\tau}{\nu}(1-\nu^2\frac{t}{t_k})\times\log\frac{1-\nu^2t/t_k}{1-\nu t/t_k}$	$1+\frac{\theta}{k}$
0	5	1.0000	0	-.0576	215.0	215.4	400	-161	-2	237	0	0	34680	-336668	1.0111
5	5	.9440	-.0576	-.0612	215.8	217.2	428	-161	-10	257	0	237	36155	-339931	1.0591
10	5	.8880	-.1188	-.0651	218.6	220.8	462	-161	-26	275	0	494	38020	-343712	1.1602
15	5	.8320	-.1839	-.0697	223.0	225.9	507	-161	-64	282	0	769	40215	-351722	1.3295
20	5	.7760	-.2536	-.0749	228.8	232.2	559	-161	-98	300	0	1051	42640	-355996	1.3795
26	5	.7200	-.3285	-.0810	235.6	239.0	623	-161	-106	357	19	1332	45235	-360951	1.6237
30	5	.6640	-.4095	-.0882	242.4	245.7	698	-161	-103	434	39	1669	48100	-371824	1.6672
35	5	.6079	-.4977	-.0967	249.0	251.9	785	-161	-161	526	59	2083	51265	-376445	1.6390
40	5	.5519	-.5944	-.1070	254.8	257.0	886	-161	-161	643	72	2590	54715	-391919	1.6067
45	5	.4959	-.7014	-.0971	259.2	260.1	813	-132	-82	635	95	3216	47913	-326670	1.5002
49.1	4.1	.4500	-.7985		261.0				-46		104	3842			1.3570

WORK SHEET FORM 30-103-1 (REV. 1-43)

1st BURNING PERIOD
$\nu = .55 \qquad \alpha = 15°$

δt	$\frac{(\delta t)^2}{2}(1+\frac{\tilde e}{w})$	δh	Correction due to θ	h	p	Co	$\frac{pCo(\%/1000)^2}{1-\nu^2 \eta}$	D/w	b/w
0			0	0	0	0	0	0	0
5	-407	605	0	605	2080	.3	37.129	.0222	.0111
10	-427	1777	0	2382	1970	.3	162.42	.0972	.0597
15	-467	3175	0	5557	1750	.3	373.15	.2232	.1602
20	-559	4534	0	10091	1450	.45	928.80	.5558	.3895
25	-654	5990	850	15231	1140	.4	1156.0	.6916	.6237
30	-671	7613	1749	21945	810	.3033	1074.3	.6428	.6672
35	-660	9556	2644	30606	530	.2654	1061.7	.6352	.6390
40	-647	11973	3498	41725	310	.24215	968.2	.5792	.6072
45	-604	14912	4258	55277	140	.2274	703.8	.4212	.5002
49.1	-367	14876	4664	70347	64.5	.21927	489.33	.2928	.3570

1st Burning Period

$\gamma = .55 \qquad d = 15°$

t	$\bar\theta$	$\cos\bar\theta$	$-\frac{g't}{V}\cos\bar\theta$	$\frac{\bar V\cos\bar\theta}{R}\delta t$, $\frac{2V_e}{R}\delta t$, δt	$\log(1-\gamma\frac{t}{Ig})$	$\log\frac{(1-\gamma t/Ig)}{(1-\gamma' t/Ig)}$	$\frac{\bar Ig\sin\theta}{V}$	$\log\frac{(1-\gamma t/Ig)}{V}$, $x\bar Ig\sin\theta$	$d\theta$	θ_{avg}	$\theta°$
24.55	1.5755	−.0047	.0001	0	.0001	−32.16				1.571	90.00
25	1.5156	.0552	−.0058	0	−.0069	−.3285	−.0810	1.4272	−.0098	−.010	1.561	84.45
30	1.4072	.1629	−.0136	0	−.0007	−.4095	−.0882	1.3020	−.1055	−.110	1.451	83.14
35	1.3031	.2645	−.0177	0	−.0007	−.4977	−.0938	1.0636	−.107	−.107	1.344	77.03
40	1.2050	.3577	−.0192	.0002	−.0007	−.5944	−.0967	.7128	−.0844	−.101	1.243	71.22
45	1.1211	.4345	−.0158	.0003	−.0005	−.7014	−.1070	.7163	−.0766	−.095	1.148	65.79
49.1			−.0003	.0005		−.7985	−.0971	.5974	−.0580	−.073	1.075	61.60

WORK SHEET FORM 30.103 (REV. 1-43)

2ND BURNING PERIOD $\nu = .55 \qquad \alpha = 15°$

INTERVALS	.7	$1-\nu^{7/10}$	$\log(1-\nu^{7/10})$	$\log\frac{1-\nu^{7/10}}{1-\nu\cdot\nu^{7/10}}$	$-g\pi\cos\alpha \times \log\frac{1-\nu}{1-\nu\cdot\nu^{7/10}}$	θ	$\sin\theta$	$-gst\sin\theta$	p	c_0	$\frac{pc_0}{1-\nu^{7/10}}$	δ/w	δ/w	$-14\,\delta/w$	δv	v
0	0	1.0000	0				.8841									.3842
1	5.085	.9450	-.0566	-.0566	.493	1.0586	.8715	-143.	64.5	.2193	206.78	.1266	.103	-18	332	4002
2	10.170	.8900	-.1165	-.0599	.522	1.00712	.8453	-138.	34.0	.2173	135.90	.080	.0645	-10	374	4174
3	15.255	.8350	-.1803	-.0638	.556	.9571	.8316	-134.	16.5	.2145	82.27	.049	.040	-6	416	4362
4	20.340	.7800	-.2485	-.0672	.594	.9080	.7883	-129.	8.5	.2122	53.23	.031	.027	-4	462	4548
5	25.425	.7250	-.3216	-.0731	.637	.8601	.7518	-124.	4.7	.2102	37.29	.023	.019	-3	510	4756
6	30.510	.6700	-.4005	-.0729	.687	.8135	.7267	-119.	2.27	.2015	23.01	.014	.011	-2	566	4964
7	35.595	.6150	-.4861	-.0856	.745	.7621	.6947	-114.						-1	630	5195
8	40.680	.5600	-.5798	-.0937	.816	.7240	.6624	-109.						0	707	5426
9	45.765	.5050	-.6832	-.1034	.901	.6809	.6295	-103.							792	5681
10	50.850	.4500	-.7985	-.1153	1.004	.6388	.5962	-98.							906	5936
						.5962	.5794									6219

Additional v values: 6502, 6817, 7132, 7426, 7839, 8238, 8637, 9090, 9543

WORK SHEET
FORM 30-103-1 (REV. 1-43)

2ND BURNING PERIOD
$V = .55$ $d = 15°$

INTERVALS	Cos θ	$-\dfrac{g\,\partial t\cos\theta}{v}$	$\dfrac{v\cos\theta}{R}$ st	$\dfrac{2Vg}{R}$ st	$\dfrac{9\sum smt}{v}$ $\times \log\!\left(\dfrac{1-v^{\frac12}t/4}{1-v^{\frac12}t/4n}\right)$	$d\theta$	$\theta_{e.n}$	$\theta°$	$V\sin\theta$	dh	h
0	.4904	-.0200	.0004	.0007	-.0330	-.0520	1.0846	62.143	3396		103477
1	.5342	-.0200	.0006	.0007	-.0320	-.0507	1.0326	59.16	3584		117747
2	.5760	-.0198	.0007	.0007	-.0313	-.0497	.9819	56.26	3782		88094
3	.6153	-.0194	.0008	.0007	-.0306	-.0485	.9322	53.41	3986	18728	106823
4	.6524	-.0188	.0009	.0007	-.0300	-.0472	.8837	50.63	4194	19750	126572
5	.6870	-.0181	.0010	.0007	-.0296	-.0460	.8365	47.93	4406	22949	147370
6	.7193	-.0173	.0012	.0007	-.0293	-.0447	.7905	45.29	4620	25181	169236
7	.7492	-.0164	.0013	.0007	-.0292	-.0436	.7457	42.73	4840		192185
8	.7770	-.0155	.0015	.0007	-.0293	-.0426	.7022	40.23	5063		216237
9	.8028	-.0145	.0018	.0007	-.0296	-.0416	.6596	37.79	5275	27469	240652
10							.6180	35.41	5529		295171

WORK SHEET
FORM 30-103-1 (REV. 1-43)

3rd BURNING PERIOD
$$\nu = .55 \qquad \alpha = 15°$$

SAME VALUES AS IN 2ND BURNING PERIOD

INTERVALS	t	$1-\nu^{3/4}/t_b$	$\log(1-\nu^{3/4}/t_b)$	$\log\dfrac{1-\nu^{3/4}/t_b}{1-\nu^{3/4}/t_b}$	$x\log\dfrac{1-\nu^{3/4}/t_b}{}$	Sin θ	$-g\sin\theta$	dv	V	V sin θ	dh	h
0						.5794	-93	+00	9543	5529	28191	295171
						.5693			9743	5544		
1						.5590	-89	433	9943	5559	28334	323362
						.5438			10159	5572		
2						.5384	-87	469	10376	5586	28476	351696
						.5280			10611	5600		
3						.5177	-83	511	10845	5614	28629	380172
						.5074			11101	5630		
4						.4971	-80	557	11356	5646	28776	408201
						.4863			11635	5659		
5						.4762	-76	611	11913	5672	28913	437577
						.4657			12219	5686		
6						.4552	-73	672	12524	5700	29061	466490
						.4447			12860	5715		
7						.4342	-69	747	13196	5730	29208	495551
						.4235			13570	5744		
8						.4130	-66	835	13943	5758	29351	524759
						.4024			14361	5772		
9						.3915	-62	942	14778	5786	29493	554110
						.3807			15249	5800		
10						.3699			15720	5814		583603

3ʳᵈ BURNING PERIOD

$V = .55$ $d = 15°$

INTERVALS	Cos θ	$-\dfrac{9S\tau\cos\theta}{V}$	$\dfrac{V\cos\theta}{RS\tau}$	$\dfrac{2V\varepsilon}{RS\tau}$	$\dfrac{9\tau\sin\tau}{V}\log\dfrac{1-\nu^2 t/\sigma}{1-\nu^2 t/\sigma}$	dθ	θ rad.	θ°
0	.8222	-.0138	.0019	.0007	-.0136	-.0248	.6180	35.41
1	.8392	-.0135	.0020	.0007	-.0137	-.0247	.5932	33.99
2	.8493	-.0131	.0021	.0007	-.0140	-.0243	.5685	32.57
3	.8617	-.0127	.0023	.0007	-.0143	-.0240	.5442	31.18
4	.8735	-.0123	.0024	.0007	-.0147	-.0239	.5202	29.81
5	.8850	-.0119	.0026	.0007	-.0151	-.0237	.4963	28.44
6	.8957	-.0114	.0027	.0007	-.0155	-.0235	.4726	27.02
7	.9059	-.0109	.0029	.0007	-.0161	-.0234	.4491	25.73
8	.9155	-.0105	.0032	.0007	-.0168	-.0234	.4257	24.39
9	.9247	-.0099	.0034	.0007	-.0176	-.0234	.4023	23.05
10							.3789	21.71

WORK SHEET
(FORM 30-103-I (REV. 1-43))

4TH BURNING PERIOD
ν = .55   α = 30°

t	ν t/76	1-ν t/76	log(1-ν t/76)	log $\frac{1-\nu^{t}/76}{1-\nu^{t+1}/76}$ / x log $\frac{1-\nu^{t}/76}{1-\nu^{t+1}/76}$	Sin θ	-g St Sinθ	δV	V	V sinθ	δh	h
45.6	.55	.45	-.798	-.476 / .3710	0	-.36	24560	0	0	0	0
22.8	.275	.725	-.323	-.322 / .0985	.0494	36.74	20226	1100	25100	25100	
0	0	1.000	-.322	.2508 / .1413	.0985	-.104	2404	2059	2059	64000	
		0	0	/ .1840	.1413	2404	12482	2800	64000	89100	
					.1840			3400	3400		

WORK SHEET
FORM 30-103-?1,(REV. 1-43)

4TH BURNING PERIOD
$\nu = .55 \qquad d = 30°$

t	$\cos\theta$	$-\dfrac{2\cos\theta}{V}$	$\dfrac{V}{R}\cos\theta$	$-\dfrac{2\cos\theta}{V}$	$SH\left(-\dfrac{2\cos\theta}{V}\right)$	$d\theta$	θ RAD.	$\theta°$
45.6	.999	-.00141	.00107	-.00019	-.0044	-.0987	0	0
22.8		-.00162	.00092	-.00055	-.0125	-.0987	.0987	5.7
0	.990	-.00162	.00092	-.00055	-.0737	-.0862	.1849	10.6

PLATE I

PERSPECTIVE CUTAWAY

OXYGEN ALCOHOL ROCKET PROPOSAL

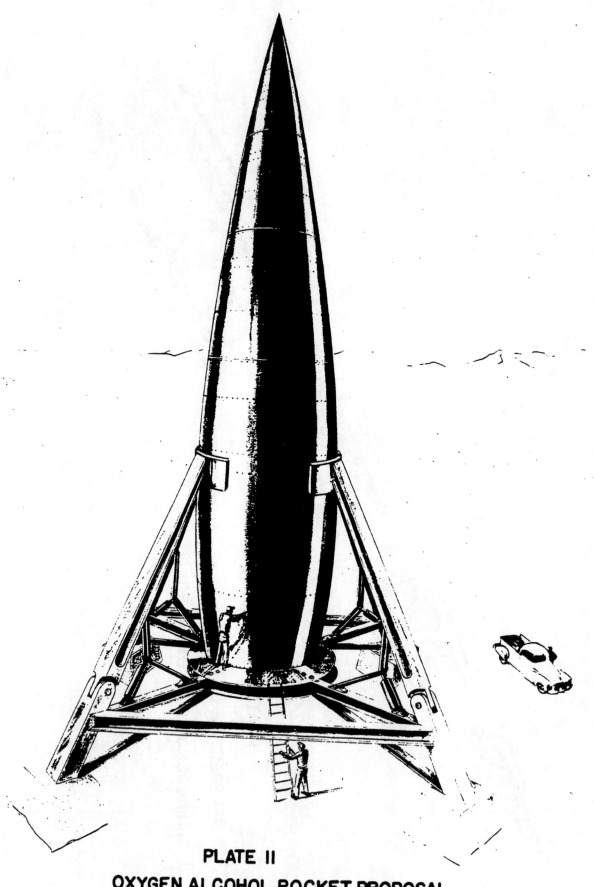

PLATE II
OXYGEN ALCOHOL ROCKET PROPOSAL
ASSEMBLED FOR FIRING

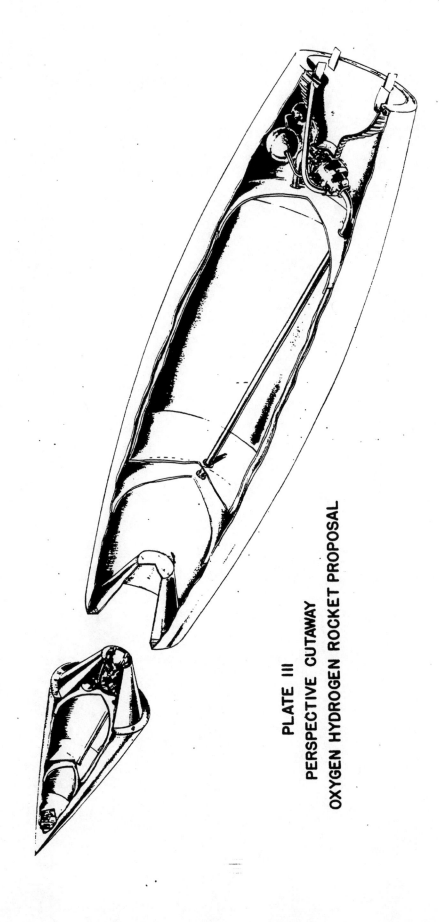

PLATE III
PERSPECTIVE CUTAWAY
OXYGEN HYDROGEN ROCKET PROPOSAL

APPENDIX E

E. Development of Small Perturbation Equations of Motion.

From Appendix C page 2 we have the equations

$$m\ddot{r} - mr(\dot{\phi}+\Omega)^2 + \frac{k\,m\,M}{r^2} = F_r$$

$$2mr\dot{r}(\dot{\phi}+\Omega) + mr^2\ddot{\phi} = rF_\phi$$

Referring to a stationary coordinate system the equations become

$$r^2\ddot{r} - r^3\dot{\phi}^2 + kM = \frac{F_r\,r^2}{m}$$

$$2\dot{r}\dot{\phi} + r\ddot{\phi} = \frac{F_\phi}{m}$$

Let $r = r_0 + \Delta r$ and $\dot{\phi} = \omega_0 + \Delta\omega$

Then $\left[r_0^2 + 2r_0(\Delta r) + (\Delta r)^2\right](\ddot{\Delta r}) - \left[r_0^3 + 3r_0^2(\Delta r) + 3r_0(\Delta r)^2 + (\Delta r)^3\right]\left[\omega_0^2 + 2\omega_0(\Delta\omega) + (\Delta\omega)^2\right]$

$+ kM = \frac{F_r}{m}\left[r_0^2 + 2r_0(\Delta r) + (\Delta r)^2\right]$

And $2(\dot{\Delta r})\left[\omega_0 + (\Delta\omega)\right] + \left[r_0 + (\Delta r)\right](\dot{\Delta\omega}) = \frac{F_\phi}{m}$

$r_0^2(\ddot{\Delta r}) - r_0^3\omega_0^2 + kM - \frac{F_r}{m}r_0^2 + (\Delta r)\left[2r_0(\ddot{\Delta r}) - 3r_0^2\omega_0^2 - \frac{2r_0 F_r}{m}\right]$

$+ (\Delta\omega)\left[-2\omega_0 r_0^3\right] \doteq 0$ Neglecting higher order terms

And $2(\dot{\Delta r})\omega_0 + r_0(\dot{\Delta\omega}) \doteq \frac{F_\phi}{m}$

If the departure from a circular orbit is small the coefficient of Δr in the first equation becomes $-3r_0^2\omega_0^2$.
Also $kM = g_0 r_0^2$.

$$(\ddot{\Delta r}) + (g_0 - r_0\omega_0^2) - \frac{F_r}{m} - 3\omega_0^2(\Delta r) - 2r_0\omega_0(\Delta\omega) \doteq 0$$

$$2(\dot{\Delta r})\omega_0 + r_0(\dot{\Delta\omega}) - \frac{F_\phi}{m} \doteq 0$$

To eliminate $(\Delta\omega)$

$$(\Delta\omega) = \frac{(\ddot{\Delta r}) + (g_0 - r_0\omega_0^2) - \frac{F_r}{m} - 3\omega_0^2(\Delta r)}{2r_0\omega_0}$$

$$(\Delta'\omega) = \frac{(\dddot{\Delta r}) - \left(\frac{\dot{F_r}}{m}\right) - 3\omega_o^2(\dot{\Delta r})}{2r_o\,\omega_o}$$

$$2(\dot{\Delta r})\omega_o + \frac{(\dddot{\Delta r}) - \left(\frac{\dot{F_r}}{m}\right) - 3\omega_o^2(\dot{\Delta r})}{2\omega_o} - \frac{F_\phi}{m} \doteq 0$$

$$(\dddot{\Delta r}) + (\dot{\Delta r})\omega_o^2 - \left(\frac{\dot{F_r}}{m}\right) - \frac{F_\phi}{m}2\omega_o \doteq 0$$

OR $\quad \dfrac{d^2 V_R}{dt^2} + \omega_o^2 V_R = \dfrac{2\omega_o F_\phi}{m} + \dfrac{d}{dt}\left(\dfrac{F_r}{m}\right) \quad$ WHERE $V_R = (\dot{\Delta r})$

LET $\quad V_R = A\cos(\omega_o t - \delta) + B \qquad$ (NEGLECTING $\dfrac{d}{dt}\left(\dfrac{F_r}{m}\right)$ AND

THEN $\quad B = \dfrac{2F_\phi}{\omega_o m} \qquad\qquad$ ASSUMING $\dfrac{F_\phi}{m}$ CONSTANT)

$$V_R = A\cos(\omega_o t - \delta) + \frac{2F_\phi}{\omega_o m}$$

$$\Delta r = A\int_0^t \cos(\omega_o t - \delta)dt + \frac{2}{\omega_o}\int_0^t \frac{F_\phi}{m}\,dt$$

$$\left(\frac{\Delta r}{r_o}\right) \text{ PER REVOLUTION} = \frac{A}{r_o}\int_0^{\frac{2\pi}{\omega_o}} \cos(\omega_o t - \delta)dt + \frac{2}{\omega_o r_o}\int_0^{\frac{2\pi}{\omega_o}} \frac{F_\phi}{m}\,dt$$

$$\left(\frac{\Delta r}{r_o}\right) \text{ PER REVOLUTION} = 4\pi\left(\frac{F_\phi}{W}\right)\frac{g_o}{r_o\,\omega_o^2} \quad \text{FOR } \frac{F_\phi}{m} \text{ CONSTANT}$$

BUT $\quad g_o \doteq r_o\,\omega_o^2$

AND $\left(\dfrac{\Delta r}{r_o}\right)$ PER REVOLUTION $= 4\pi\left(\dfrac{F_\phi}{W}\right)$

Appendix F

F. ORBIT CALCULATION

Orbital Motion Under the Newtonian Law* - In this appendix, the equations of motion of a body will be developed in a form suitable for use in calculating the trajectory of a body after it has been accelerated to the proper speed and direction for orbital motion.

The body is treated as a particle of unit mass acting under a central force varying as the inverse square of the distance from the center of the earth, in accordance with the Newtonian law of universal gravitation.

If (r, ϕ) be the coordinates of the body with respect to the central force, the kinetic energy of the particle is

$$T = 1/2 \; (\dot{r}^2 + r^2 \dot{\phi}^2).$$

Letting P denote the acceleration directed to the center of force, the work done by the force in an arbitrary infinitesimal displacement, $(dr, d\phi)$ is equal to $-Pdr$.

The Lagrangian equations of motion for the particle are

$$\begin{cases} \ddot{r} - r\dot{\phi}^2 = -P, \\ \dfrac{d(r^2\dot{\phi})}{dt} = 0. \end{cases}$$

The latter of these equations gives on integration

$$r^2\dot{\phi} = H,$$

where H is a constant. This integral corresponds to the ignorable coordinate ϕ, and can be interpreted physically as the integral of angular momentum of the particle about the center of force.

Eliminating dt from the first equation by use of the relationship

* Whittaker, "A Treatise on the Analytical Dynamics of Particles and Rigid Bodies", Cambridge 1937; PP. 86-90.

Appendix F

$$\frac{d}{dt} = \frac{H}{r^2} \frac{d}{d\phi} \quad \text{we obtain}$$

$$\frac{H}{r^2} \frac{d}{d\phi}\left(\frac{H}{r^2}\frac{dr}{d\phi}\right) - \frac{H^2}{r^3} = -P.$$

Letting $u = \frac{1}{r}$, we obtain the differential equation of the path,

$$\frac{d^2u}{d\phi^2} + u = \frac{P}{H^2u^2} . \tag{1}$$

If the particle be projected from the point whose polar coordinates are (R_o, α) with a velocity V_s in a direction making an angle γ with R_o, the angular momentum is

$$H = R_o V_s \sin \gamma .$$

If the central force per unit mass be μu^2, then

$$P = \mu u^2 .$$

Substituting this relation in equation (1) we have

$$\frac{d^2u}{d\phi^2} + u = \frac{\mu}{V_s^2 R_o^2 \sin^2\gamma}$$

This is a linear differential equation with constant coefficients whose integral is

$$u = \frac{\mu}{V_s^2 R_o^2 \sin^2\gamma} \left\{ 1 + e \cos (\phi - \overline{w}) \right\},$$

where e and $\overline{w}$ are constants of integration. This is the equation of a conic in polar coordinates whose focus is at the origin, whose eccentricity is e, and whose semi-latus rectum is

$$l = \frac{V_s^2 R_o^2 \sin^2\gamma}{\mu}$$

The constant $\overline{w}$ determines the position of the apse-line.

Appendix F

Initially we have

$$\phi = \alpha, \quad u = \frac{1}{R_o}, \quad \text{and} \quad \frac{du}{d\phi} = -\frac{\cot \gamma}{R_o}.$$

Hence, it follows that

$$e^2 = 1 + \frac{V_s^4 R_o^2 \sin^2 \gamma}{\mu^2} - 2\frac{V_s^2 R_o \sin^2 \gamma}{\mu}, \quad \text{and}$$

$$\cot (\alpha - \bar{w}) = \frac{-\mu}{R_o V_s^2 \sin \gamma \cos \gamma} + \tan \gamma.$$

The semi-major axis, when conic is an ellipse is generally denoted

by a, and is given by

$$a = \frac{\ell}{1 - e^2}.$$

Substituting the values of e^2 and ℓ already determined, we have

$$a = \frac{R_o}{2 - \frac{V_s^2 R_o}{\mu}}$$

which determines a in terms of the initial data.

If V_o be the orbital velocity of the particle, then by equating

central forces,

$$\frac{\mu m}{r^2} = \frac{m V_o^2}{r}.$$

For $r = R_o$, this gives $V_o^2 = \frac{\mu}{R_o}$. Defining V_s by the relation

$$V_s = V_o \left(1 + \frac{\Delta V}{V_o}\right), \quad \text{we obtain}$$

$$\frac{V_s^2 R_o}{\mu} = \left(1 + \frac{\Delta V}{V_o}\right)^2.$$

Appendix **F**

Making this substitution in the equations previously derived, we have

$$a = \frac{R_o}{2 - \left(1 + \frac{\Delta V^2}{V_o}\right)} ,$$

$$e^2 = 1 - \sin^2 \gamma + 4 \left(\frac{\Delta V^2}{V_o}\right) \sin^2 \gamma \left[1 + 1/2 \frac{\Delta V}{V_o}\right]^2 , \quad \text{and}$$

$$\ell = R_o \sin^2 \gamma \left(1 + \frac{\Delta V^2}{V_o}\right) .$$

If R_{min} be the minimum radius of the orbit from the center of the earth,

$$e = 1 - \frac{R_{min}}{a} = 1 - \frac{R_{min}}{R_o} \left[2 - \left(1 + \frac{\Delta V}{V_o}\right)^2\right] .$$

It will be useful to know the difference between the maximum and minimum distance of the orbit above the earth's surface. Denoting this quantity by ΔH_{max}, we have

$$\Delta H_{max} = 2ae$$

$$= 2R_{min} \left(\frac{e}{1 - e}\right)$$

$$= 2 \left[\frac{R_o}{2 - \left(1 + \frac{\Delta V}{V_o}\right)^2} - R_{min}\right] .$$

If we wish to know the required height at the start of the orbit for given values of ΔH_{max}, R_{min}, and $\frac{\Delta V}{V_o}$, we find

$$R_o = \left[R_{min} + \frac{\Delta H_{max}}{2}\right] \left[2 - \left(1 + \frac{\Delta V}{V_o}\right)^2\right]$$

Letting $\Delta H = R_o - R_{min}$, denote the maximum altitude loss with respect to the starting altitude, we get

$$\Delta H = \left[R_{min} + \Delta H_{max}\right] - \left[R_{min} + \frac{\Delta H_{max}}{2}\right] \left[1 + \frac{\Delta V}{V_o}\right]^2 .$$

Appendix F

The following pages contain plots of the relationships between the major characteristics of the orbit.

Analysis **ORBIT VEHICLE**

Prepared by **C. V. STURDEVANT**

Date **4-26-46**

S.M. Plant

Page **6 F**

Model **#1033**

Report No. **SM 11827**

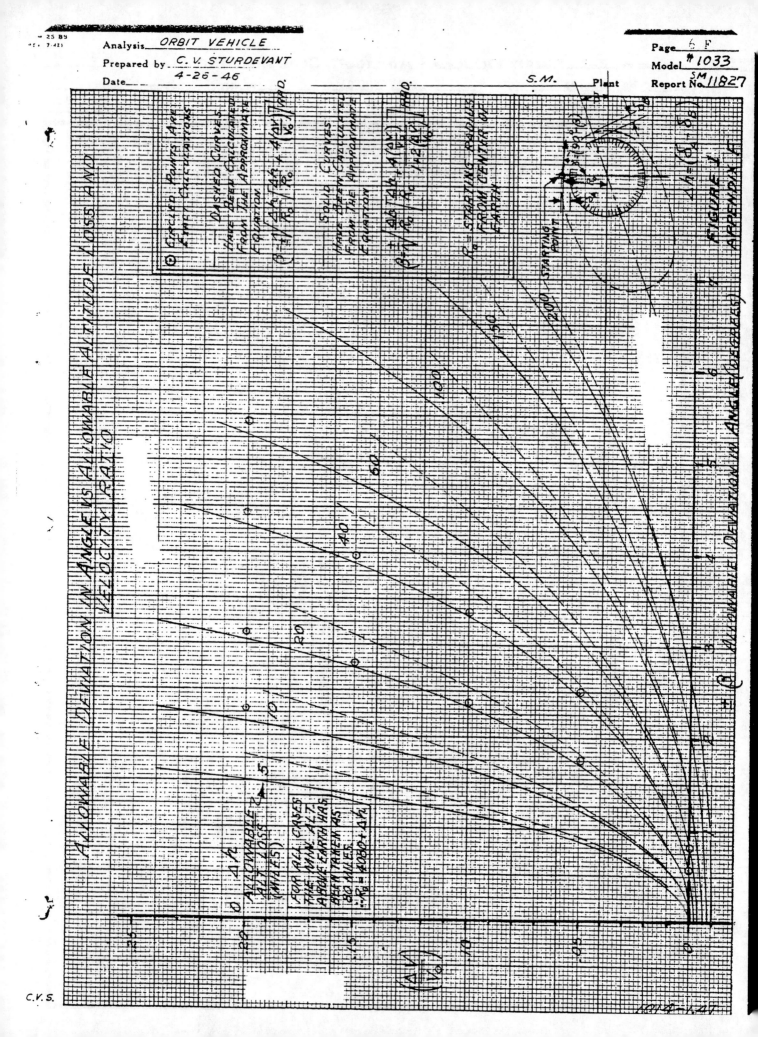

FIGURE 1

APPENDIX F

$\Delta h = (\beta + 1)(R_o - R_g)$

$(+) \Delta h =$ ALLOWABLE DEVIATION IN ALTITUDE (MILES)

ALLOWABLE DEVIATION IN ANGLE VS ALLOWABLE ALTITUDE LOSS AND VELOCITY RATIO

$-(+ \alpha =$ ALLOWABLE DEVIATION IN ANGLE (DEGREES)

$\left(\frac{\Delta V}{V_o}\right) =$ ALLOWABLE VELOCITY RATIO

⊕ CIRCLED POINTS ARE FIRST CALCULATIONS

― ― ― DASHED CURVES HAVE BEEN CALCULATED FROM THE APPROXIMATE EQUATION

$(\beta + 1) \left[\frac{\Delta h}{R_o} + 4 \left(\frac{\Delta V}{V_o}\right) \right] / \text{RAD.}$

SOLID CURVES HAVE BEEN CALCULATED FROM THE APPROXIMATE EQUATION

$(\beta + 1) \sqrt{\frac{R_g}{R_o} \left[\frac{\Delta h}{R_o} + 1 \left(\frac{\Delta V}{V_o}\right) \right] } / \text{RAD.}$

$R_o =$ STARTING RADIUS FROM CENTER OF EARTH

FOR ALL CASES THE MIN. ALT. ABOVE EARTH HAS BEEN TAKEN AS 80 MILES. $R_g = 4050 + \frac{1}{2}$

ALLOWABLE ALT. LOSS (MILES)

200 STARTING POINT

C.V.S.

ELLIPSES   HYPERBOLAS

PARABOLAS

ORIENTATION OF ELLIPSES

ORBIT VEHICLE

REF. 1014-1.33, 1.34

$\frac{\Delta V}{V_0}$ ~ VELOCITY INCREASE OVER ORBITAL AT LAUNCHING ALTITUDE.

γ = ANGLE OF FLIGHT PATH AT STARTING ALTITUDE WITH THE VERTICAL = $90° - \beta$

ω = ANGLE THE MAJOR AXIS OF THE ELLIPSE MAKES WITH THE HORIZONTAL STARTING POINT ASSUMED TO BE VERTICALLY ABOVE THE EARTH

$$\tan \omega = \tan \gamma \left[1 - \frac{1}{\left(1 + \frac{\Delta V}{V_0}\right)^2 \tan^2 \beta} \right]$$

FOR $\gamma > 90°$ (i. e. β NEGATIVE)

$\omega = (180°)$ MINUS (ω FOR β POSITIVE)

$\left(\frac{\Delta V}{V_0}\right)$ FIGURE 2

APPENDIX F

1014-1.35

C.V.S.

M 25 BS
(REV. 7-42)

Analysis __ORBIT VEHICLE__

Prepared by __C.V. STURDEVANT__

Date _____4-23-46_____

Page __8 F__

Model __#1033__

Plant _____

Report No. __SM 1182?__

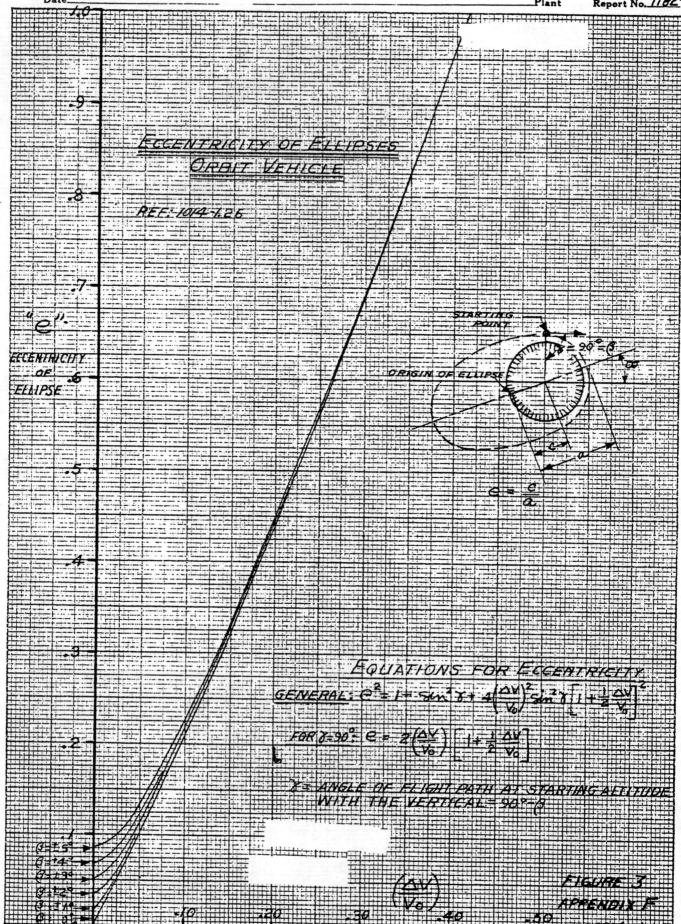

ECCENTRICITY OF ELLIPSES
ORBIT VEHICLE

REF: 1014-7.26

"e"

ECCENTRICITY
OF
ELLIPSE

STARTING POINT

ORIGIN OF ELLIPSE

$\delta = 90° - \beta$

$e = \dfrac{c}{a}$

EQUATIONS FOR ECCENTRICITY

GENERAL: $e^2 = 1 - \sin^2\gamma + 4\left(\dfrac{\Delta V}{V_0}\right)^2 \sin^2\gamma \left[1 + \dfrac{1}{2}\dfrac{\Delta V}{V_0}\right]^2$

FOR $\gamma = 90°$: $e = 2\left(\dfrac{\Delta V}{V_0}\right)\left[1 + \dfrac{1}{2}\dfrac{\Delta V}{V_0}\right]$

$\gamma =$ ANGLE OF FLIGHT PATH AT STARTING ALTITUDE WITH THE VERTICAL $= 90° - \beta$

$\left(\dfrac{\Delta V}{V_0}\right)$

FIGURE 3

APPENDIX F

1014-7.217

C.V.S.
M.G.

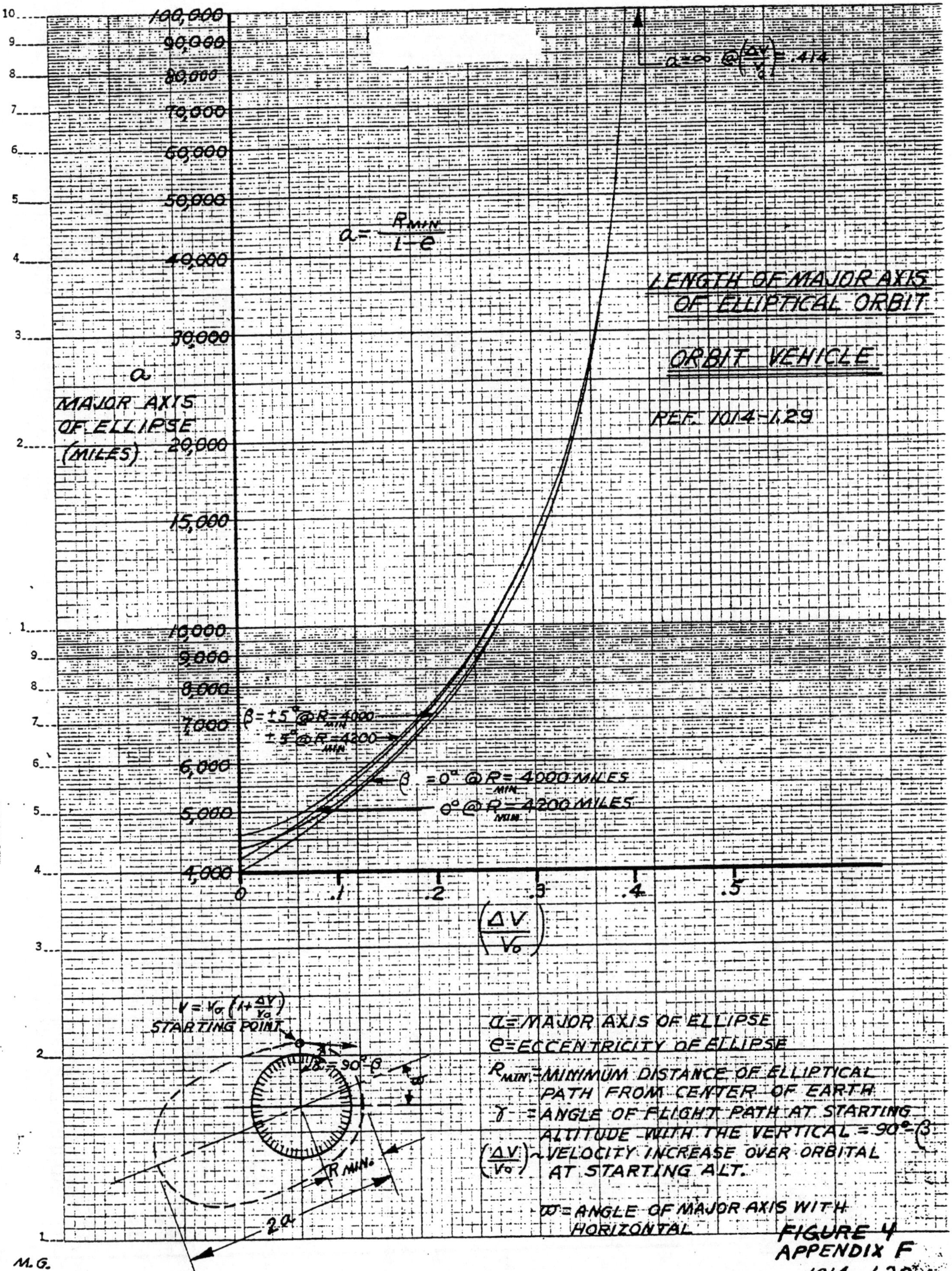
100,000
90,000
80,000
70,000
60,000
50,000
40,000
30,000
20,000
15,000
10,000
9,000
8,000
7,000
6,000
5,000
4,000
a
MAJOR AXIS OF ELLIPSE (MILES)
a = R_MIN / 1-e
a=∞ @ (ΔV/V₀) = .414
LENGTH OF MAJOR AXIS OF ELLIPTICAL ORBIT
ORBIT VEHICLE
REF. 1014-1.29
β = ±5° @ R_MIN = 4000
±5° @ R_MIN = 4200
β = 0° @ R_MIN = 4000 MILES
0° @ R_MIN = 4200 MILES
0
.1
.2
.3
.4
.5
(ΔV / V₀)
V = V₀ (1 + ΔV/V₀)
STARTING POINT
γ = 90° - β
ϖ
R MIN.
2a
a = MAJOR AXIS OF ELLIPSE
e = ECCENTRICITY OF ELLIPSE
R_MIN = MINIMUM DISTANCE OF ELLIPTICAL PATH FROM CENTER OF EARTH
γ = ANGLE OF FLIGHT PATH AT STARTING ALTITUDE WITH THE VERTICAL = 90° - β
(ΔV/V₀) ~ VELOCITY INCREASE OVER ORBITAL AT STARTING ALT.
ϖ = ANGLE OF MAJOR AXIS WITH HORIZONTAL
M.G.

FIGURE 4
APPENDIX F
1014-1.30

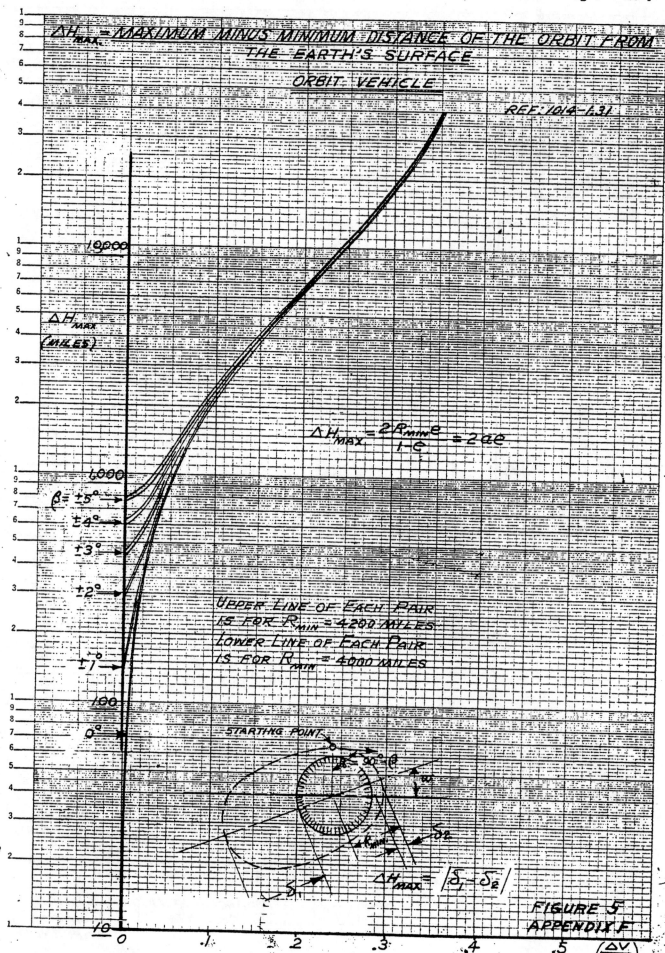

$\Delta H_{MAX.}$ = MAXIMUM MINUS MINIMUM DISTANCE OF THE ORBIT FROM THE EARTH'S SURFACE

ORBIT VEHICLE

REF: 1014-1.31

$\Delta H_{MAX.} = \dfrac{2R_{MIN} e}{1-e} \doteq 2ae$

$\beta = \pm 5°$
$\pm 4°$
$\pm 3°$
$\pm 2°$
$\pm 1°$
$0°$

UPPER LINE OF EACH PAIR IS FOR $R_{MIN} = 4200$ MILES

LOWER LINE OF EACH PAIR IS FOR $R_{MIN} = 4000$ MILES

STARTING POINT

$\Delta H_{MAX} = |\delta_1 - \delta_2|$

FIGURE 5
APPENDIX F

KEUFFEL & ESSER CO., N. Y. NO. 359-91
Semi-Logarithmic, 4 Cycles × 10 to the inch,
MADE IN U. S. A.

Analysis ORBIT VEHICLE

Prepared by C. V. STURDEVANT

Date 5-1-46

SANTA MONICA Plant

Page 11 F

Model #1033

Report No. SM 11827

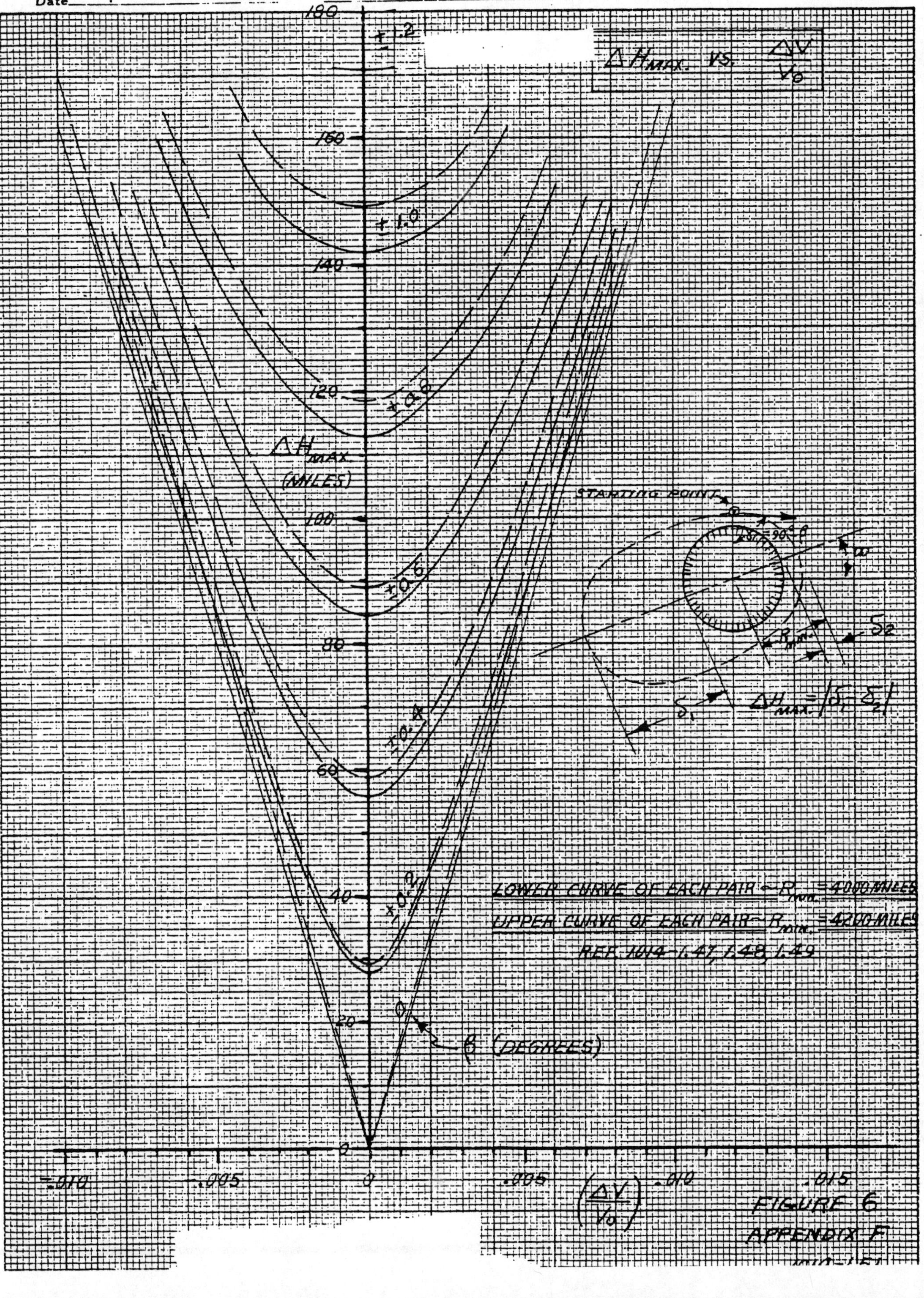

$\Delta H_{MAX.}$ VS. $\frac{\Delta V}{V_0}$
ΔH_{MAX} (MILES)
180
160
140
120
100
80
60
40
20
0
±1.2
±1.0
±0.8
±0.6
±0.4
±0.2
0
β (DEGREES)
STARTING POINT
90°-β
R_{MIN}
δ_1
δ_2
ω
$\Delta H_{MAX.} = |\delta_1 - \delta_2|$
LOWER CURVE OF EACH PAIR ~ $R_{MIN.}$ = 4000 MILES
UPPER CURVE OF EACH PAIR ~ $R_{MIN.}$ = 4200 MILES
REF. 1014-1.47, 1.48, 1.49
-.010
-.005
0
.005
.010
.015
$\left(\frac{\Delta V}{V_0}\right)$

FIGURE 6

APPENDIX F

Appendix F

Development of Approximate Orbital Equations:

Let R =   Radius

V =   Velocity

β =   Inclination to the horizontal

V_c =   Equilibrium Velocity in circular orbit at R_o

ΔV = $V_o - V_c$

Δh = $R_o - R_m$

δ = $\dfrac{\Delta h}{R_o}$

Subscript zero denotes initial condition

Conservation of energy.
$$\frac{V^2 - V_o^2}{2 g_o R_o} = \frac{R_o}{R} - 1$$

Conservation of Angular Momentum
$$\frac{R}{R_o} \times \frac{V}{V_o} \frac{\cos\beta}{\cos\beta_o} = 1$$

Eliminate V
$$\frac{\left(\dfrac{R_o}{R}\right)^2 \dfrac{V_o^2 \cos^2\beta_o}{\cos^2\beta} - V_o^2}{2 g_o R_o} = \frac{R_o}{R} - 1$$

R is a mininum (or maximum) when $\beta = 0$

$$\frac{\left(\dfrac{R_o}{R_m}\right)^2 V_o^2 \cos^2\beta_o - V_o^2}{2 g_o R_o} = \frac{R_o}{R_m} - 1$$

$$\left(\frac{R_o}{R_m}\right)^2 - \left(\frac{R_o}{R_m}\right)^2 \sin^2\beta_o - 1 = \frac{2 g_o R_o}{V_o^2}\left(\frac{R_o}{R_m} - 1\right)$$

Appendix F

$$\sin^2\beta_o = 1 - \left(\frac{R_m}{R_o}\right)^2 - \frac{2g_o R_o}{V_o^2} \times \frac{R_m}{R_o}\left(1 - \frac{R_m}{R_o}\right)$$

Let $\frac{R_m}{R_o} = 1 - \delta$, $\quad V_o = V_c + \Delta V$, $\quad \beta_o$ small

$$\sin\beta_o = \sqrt{1 - (1 - 2\delta + \delta^2)\ \frac{-2g_o R_o \times (1-\delta)\delta}{\left(V_c^2 + 2V_c\Delta V + \Delta V^2\right)}}$$

$$\beta_o \cong \sqrt{\delta^2 + 4\delta\frac{\Delta V}{V_c}}, \qquad \beta_o \cong \sqrt{\left(\frac{\Delta h}{R_o}\right)\left[\frac{\Delta h}{R_o} + 4\frac{\Delta V}{V_c}\right]}$$

Appendix F

EQUATIONS FOR CORRECTION OF ORBIT

Let T = Thrust

$\dot{m}$ = Mass

V = Velocity

I = Impulse

W_f = Fuel Weight

W = Gross Weight

C = Exhaust velocity

β = Inclination to the horizontal

(A) Correction of Angle by Thrust $\perp$ to Flight Path

$$T\,dt = m\,dV$$

$$I = \text{impulse} \cong \beta\, mV$$

$$\Delta W_f = \frac{gI}{C} = \frac{mV\beta g}{C} = \frac{WV\beta}{C}$$

$$\frac{\Delta W_f}{W} = \left(\frac{V}{C}\right)\beta$$

(B) Correction of Velocity by Thrust $\parallel$ to Flight Path

$$T\,dt = m\,dV$$

$$I = \text{Impulse} = m\Delta V$$

$$\Delta W_f = \frac{gI}{C} = \frac{g}{C}m\Delta V$$

$$\left(\frac{\Delta W_f}{W}\right) = \left(\frac{\Delta V}{C}\right)$$

Appendix F

(C) <u>Correction of Velocity and Angle by Thrust in One Direction</u>

$$\text{Tdt} \sin \Theta = mV$$

$$\text{Tdt} \cos \Theta = m\Delta V$$

$$\left(\frac{mV\beta}{I}\right)^2 + \left(\frac{m\Delta V}{I}\right)^2 = 1$$

$$I = \sqrt{m^2 V^2 \beta^2 + m^2 \Delta V^2} = \frac{\Delta W_f \, C}{g}$$

$$\frac{\Delta W_f}{W} = \sqrt{\left(\frac{V}{G}\right)^2 \beta^2 + \left(\frac{\Delta V}{C}\right)^2}$$

(D) <u>Correction of Angle by Aerodynamic Forces</u>

$$\text{Ldt} = mdV \qquad\qquad L = D \times \left(\frac{L}{D}\right)$$

$$\left(\frac{L}{D}\right) \times \text{Ddt} = mdV$$

Setting $D = T$

$$\left(\frac{L}{D}\right) \times \text{Tdt} = mdV$$

$$\frac{\Delta W_f}{W} = \left(\frac{V}{C}\right) \times \beta \times \left(\frac{D}{L}\right)$$

RM 25-S-1
(REV. 8-43)

PREPARED BY: G. Grimminger
D. D. Wall   DOUGLAS AIRCRAFT COMPANY, INC.   PAGE: 1 G

DATE: May 2, 1946   SANTA MONICA   PLANT   MODEL: #1033

TITLE: PRELIMINARY DESIGN OF SATELLITE VEHICLE   REPORT NO. SM-11827

Appendix G

G. THE METEORITE-HIT PROBABILITY FORMULAS

In the section of Chapter 11 dealing with the probability of a meteorite hitting a satellite vehicle, certain probabilities and probability-based time intervals were presented in the Tables. The derivation of the formulas used to compute these quantities is given below.

The meteorites entering the atmosphere will be assumed to have a random distribution both as regards their surface distribution over the atmospheric layer surrounding the earth and as regards their occurrence with time. It is assumed that the meteorites travel through the atmosphere along the vertical and that the planform area of the vehicle is normal to the vertical.

It is not difficult to see that the meteorite velocity and the vehicle velocity are not involved in the computation of the probability that a meteorite will strike the vehicle. This follows essentially from the assumption that the distribution of the meteorites is random with respect to surface area and time. Thus there will be a certain number N of meteorites of specified size which enter the atmosphere in each 24 hour period, and for any exposed area A_b, it is equally likely that this area will be hit regardless of where it may be situated on the surface of the atmospheric shell. This means that the area is equally likely to be hit whether it is moving or stationary and therefore the speed of the moving area is immaterial.

Let the unit of time be the hour and let an event be said to occur when a meteorite hits the vehicle. Then the average number of events $\bar{n}$ which occur in a unit of time (1 hour) is given by

$$\bar{n} = \frac{N A_b}{24 A_e} , \quad - - - - - - - - - - - - - - - (1)$$

ORM 25-S-1
(REV. 8-43)

PREPARED BY: G. Grimminger
D. D. Wall师

DOUGLAS AIRCRAFT COMPANY, INC.

DATE: May 2, 1946   SANTA MONICA   PLANT

TITLE: PRELIMINARY DESIGN OF SATELLITE VEHICLE

PAGE: 2 G

MODEL: #1033

REPORT NO. SM-11827

Appendix G

and the average time $\bar{t}$ required for the event to occur is the reciprocal of this, or

$$\bar{t} = \frac{1}{\bar{n}} = \frac{24A_e}{NA_b} \quad \text{-------------------- (2)}$$

It is obvious that the probability of the occurrence of an event must increase as the time t increases. The way in which the time must enter is determined by the Poisson exponential. [1],[2]   When an event happens on the average once in the time $\bar{t}$, the average number $\bar{m}$ of events in the time T is

$$\bar{m} = \frac{T}{\bar{t}} = \frac{NA_b}{24A_e} T \quad \text{----------------- (3)}$$

Then, according to the Poisson distribution, the probability p_r that the event will happen exactly r times in the time interval T is given by

$$p_r = \frac{\bar{m}^r e^{-\bar{m}}}{r!} , \quad \text{------------------ (4)}$$

where e is the exponential, e = 2.71828. Further, the probability p_1 that the event will happen <u>exactly once</u> in the time T is

$$p_1 = \bar{m}\, e^{-\bar{m}} \quad \text{----------------------- (5)}$$

The probability p_o that the event will fail to happen in the time T (i.e. for the event to happen zero times) is, from (4),

$$p_o = p(0) = e^{-\bar{m}} \quad \text{------------------- (6)}$$

(1) Kenney, J. F.: Mathematics of Statistics. D. Van Nostrand Co., New York, 1941, p. 29 Part 2.
(2) Freeman, H. A.: Industrial Statistics. John Wiley and Sons, New York, 1942, p. 149.

RM 25-S-1
REV. 6-43)

PREPARED BY: G. Grimminger / D. D. Wall

DATE: May 2, 1946

TITLE: PRELIMINARY DESIGN OF SATELLITE VEHICLE

DOUGLAS AIRCRAFT COMPANY, INC.

SANTA MONICA PLANT

PAGE: 3 G

MODEL: #1033

REPORT NO. SM-11827

Appendix G

If p_{1+} denote the probability that the event occur at least once in the time T it follows that $p_{1+} + p_0 = 1$ and therefore

$$p_{1+} = 1 - e^{-\bar{m}}. \text{- -}(7)$$

Inserting the value for $\bar{m}$, the value for this probability becomes

$$p_{1+} = 1 - \left[e\right]^{-\frac{NA_b}{24A_e}T} \text{- - - - - - - - - - - - - - -}(8)$$

This value of p_{1+} gives the probability that the vehicle will be hit at least once in T hours. It does not exclude the possibility that more than one hit will occur in this time interval, and in fact, definitely allows that more than one hit may occur. However, although p_{1+} does not specify the probability of the exact number of hits in the time T it is considered to best represent the type of probability which is most significant since, from Eq. (8), it is seen that $p_{1+} = 0$ for N or T = 0 and that p_{1+} increases as N and T increase.

The probability p_1, on the other hand, which from (5) may be written

$$p_1 = \frac{NA_b}{24A_e}T\left[e\right]^{-\frac{NA_b}{24A_e}T}, \text{- - - - - - - - - - - - - - -}(9)$$

does exclude the possibility of more than one hit and refers only to exactly one hit, no more and no less. It is evident the p_1 is a much more restricted type of probability than p_{1+} and has the odd characteristic of becoming smaller when the number N or T is very large. This of course follows from the fact that since p_1 refers only to exactly one hit, when N or T become larger and larger and there are thus more and more chances

FORM 25-S-1
(REV. 8-43)

PREPARED BY: G. Grimminger / D. D. Wall

DATE: May 2, 1946

TITLE: PRELIMINARY DESIGN OF SATELLITE VEHICLE

DOUGLAS AIRCRAFT COMPANY, INC.

SANTA MONICA PLANT

PAGE: 4 G

MODEL: #1033

REPORT NO. SM-11827

Appendix G

for a hit, the chances that the vehicle will be hit <u>only once</u> will become smaller.

The third probability of interest p_o, the probability for no hit at all, is evaluated from

$$p_o = \left[e\right]^{-\frac{NA_b}{24A_e}T} \quad \text{-------------------(10)}$$

Comparing Eqs. (8), (9), and (10), it is seen that as the time T increases, p_{1+} approaches 1, p_1 must go through a maximum, and p_o approaches 0.

Certain probability-based time intervals of interest are as follows.

(a) The time such that the vehicle has a 50 to 50 chance of not being hit. In this case $p_o = 0.5$ and the corresponding time to satisfy this condition will be denoted by T (0.5). From (10), this is evaluated from

$$T(0.5) = -\frac{24A_e}{NA_b}\log_e 0.5 = -\frac{9.96 \times 10^{13}}{N}, \quad \text{-------}(11)$$

where the value $\frac{24A_e}{A_b} = 1.437 \times 10^{14}$ is used.

(b) The time such that the vehicle has a 100 to 1 chance of not being hit. In this case $p_o = 0.99$, and denoting this time by T (0.99) it follows from (10) that

$$T(0.99) = -\frac{24A_e}{NA_b}\log_e 0.99 = -\frac{1.437 \times 10^{12}}{N}. \quad \text{-------}(12)$$

(c) The time such that the vehicle has a 1,000 to 1 chance of not being hit. In this $p_o = 0.999$ and the corresponding time T (0.999) is given by

$$T(0.999) = -\frac{24A_e}{NA_b}\log_e 0.999 = -\frac{1.437 \times 10^{11}}{N}. \quad \text{-----}(13)$$

Appendix G

Before presenting the various probabilities, a few remarks will be made concerning the number N. In computing the probabilities of a hit, one may consider either the total number of meteorites of all sizes, or only the total number of a certain size, or else the total number of a certain size plus all those of larger size. In considering the probabilities of a hit by a meteorite we are not especially concerned with the entire total number of meteorites of all possible sizes since many of these are too small to do any damage. On the other hand we are concerned with meteorites of a certain given size and especially those sizes which can cause damage and at the same time occur with considerable frequency. Furthermore, since when considering a certain given size the total number of all larger sizes may be appreciable, this suggests also the consideration of the total number of meteorites down to and including those of given size.

The number of meteorites for these two cases are given in Table 1, where meteors of magnitude less than -3 have not been included since they occur too infrequently to be of any importance. The table is based on the fact that when the magnitude increases by 5, the number increases by 10^2 and therefore for a change of 1 magnitude the number changes by a factor of 2.5. Thus, if there are N' meteors of magnitude M' in each 24 hour period, there will be

$$N = N' \times (10)^{\frac{2}{5}(M - M')} \quad ------------- \quad (14)$$

meteors of magnitude M in the same period.

FORM 25-S-1
(REV. 8-43)
PREPARED BY: G. Grimminger
D. D. Wall   DOUGLAS AIRCRAFT COMPANY, INC.   PAGE: 6 G
DATE: May 2, 1946   SANTA MONICA   PLANT   MODEL: #1033
TITLE: PRELIMINARY DESIGN OF SATELLITE VEHICLE   REPORT NO. SM-11827

Appendix G

The sum of all those of magnitude -3 up to and including magnitude M is given by

$$S_M = N' \times (10)^{-\frac{2}{5}M'} \sum_{-3}^{M} (10)^{\frac{2}{5}M} \quad \text{- - - - - - - - - -(15)}$$

Using the relation for the sum of a geometric series this reduces to the form

$$S_M = \frac{10^{\frac{2}{5}}}{10^{\frac{2}{5}} - 1} \times N' \times (10)^{\frac{2}{5}(M - M')} \left[1 - (10)^{-\frac{2}{5}(M + 4)} \right] \quad \text{- - - -(16)}$$

Choosing a meteor of magnitude 0 as a basis for the computation, $M' = 0$, $N' = 450,000$, and the numbers are then computed from the relation

$$N = 4.5 \times 10^5 \times (10)^{\frac{2M}{5}}, \text{ and } \text{- - - - - - - - - - - - - - - - -(17)}$$

$$S_M = 7.47 \times 10^5 \left[(10)^{\frac{2M}{5}} - .025 \right] \quad \text{- - - - - - - - - - - - - -(18)}$$

Comparing the values of N and S_M in Table 1, it is seen that for magnitude 10 for example the number S_M is about 66 per cent greater than N, a not inconsequential increase.

Appendix G

TABLE 1

PARTIAL AND TOTAL NUMBER OF METEORITES

Magnitude M	Number of Magnitude M (N)	Total Number from -3 Up to and Including Magnitude M (S_M)
-3	2.84×10^4	2.84×10^4
0	4.5×10^5	7.28×10^5
2	2.84×10^6	4.72×10^6
5	4.5×10^7	7.47×10^7
6	1.132×10^8	1.88×10^8
7	2.84×10^8	4.72×10^8
8	7.14×10^8	1.18×10^9
9	1.795×10^9	2.98×10^9
10	4.5×10^9	7.47×10^9
12	2.84×10^{10}	4.72×10^{10}
15	4.5×10^{11}	7.47×10^{11}
20	4.5×10^{13}	7.47×10^{13}
25	4.5×10^{15}	7.47×10^{15}
30	4.5×10^{17}	7.47×10^{17}

$$N = 4.5 \times 10^5 \times 10^{\frac{2M}{5}}, \text{ from Eq. (17).}$$

$$S_M = 7.47 \times 10^5 \left[10^{\frac{2M}{5}} - .025 \right], \text{ from Eq. (18).}$$

RM 25-5-1
REV. 6-43
PREPARED BY: E. W. Graham
DATE: May 2, 1946
TITLE: PRELIMINARY DESIGN OF SATELLITE VEHICLE

DOUGLAS AIRCRAFT COMPANY, INC.
SANTA MONICA PLANT

PAGE: 1H
MODEL: 1033
REPORT NO. SM 11827

Appendix H

H.   Development of Stability Equations

(A) Final Stage

T = thrust

m = mass

V = velocity (average)

R = radius from center of earth

M = pitching moment

I = Moment of inertia

d = displacement of thrust axis from C.G.

$$M = I\ddot{\epsilon}$$

$$T \sin \epsilon + \frac{mV^2}{R} - W = m\ddot{h} \doteq mV\dot{\theta}$$

$$\epsilon \doteq \frac{mV\dot{\theta}}{T} + \frac{W}{T} - \frac{mV^2}{RT}$$

$$LET \ \frac{M}{I} = K_0 \int \left[\theta + \Delta t_0 \dot{\theta} + \frac{(\Delta t_0)^2}{2!} \ddot{\theta} + \right] dt + K_1 \left[\theta + \Delta t_1 \dot{\theta} + \frac{(\Delta t_1)^2}{2!} \ddot{\theta} + \right]$$

$$+ K_2 \left[\dot{\theta} + \Delta t_2 \ddot{\theta} + \frac{(\Delta t_2)^2}{2!} \dddot{\theta} + \right] + K_3 \left[\epsilon + \Delta t_3 \dot{\epsilon} + \frac{(\Delta t_3)^2}{2!} \ddot{\epsilon} + \right]$$

$$+ K_4 \left[\dot{\epsilon} + \Delta t_4 \ddot{\epsilon} + \frac{(\Delta t_4)^2}{2!} \dddot{\epsilon} + \right] + \frac{Td}{I} = \ddot{\epsilon}$$

Where K_1, K_2 etc represent magnitudes of artificially applied restoring and damping moments and K_0 is an integral term necessary for approaching the correct flight path angle when eccentric thrust is present. The terms Δt_0, Δt_1 etc correspond to time lags in application of these moments.

Differentiating the equation and eliminating ϵ

$$K_0\left[\theta + \Delta t_0\,\dot\theta + \frac{\Delta t_0^2}{2!}\ddot\theta + \right] + K_1\left[\dot\theta + \Delta t_1\,\ddot\theta + \frac{\Delta t_1^2}{2!}\dddot\theta + \right]$$

$$+ K_2\left[\ddot\theta + \Delta t_2\,\dddot\theta + \frac{\Delta t_2^2}{2!}\ddddot\theta + \right] + K_3\frac{mV}{T}\left[\ddot\theta + \Delta t_3\,\dddot\theta + \frac{\Delta t_3^2}{2!}\ddddot\theta + \right]$$

$$+ K_4\frac{mV}{T}\left[\dddot\theta + \Delta t_4\,\ddddot\theta + \frac{\Delta t_4^2}{2!}\dddddot\theta + \right] - \frac{mV}{T}\ddddot\theta = 0$$

For small time lags the equation becomes:

$$\ddddot\theta + \dddot\theta\left[-K_4\right] + \ddot\theta\left[-K_3 - \frac{K_2 T}{mV}\right] + \dot\theta\left[-\frac{K_1 T}{mV}\right] + \theta\left[-\frac{K_0 T}{mV}\right] = 0$$

The conditions for stability are that all coefficients of θ, $\dot\theta$ etc must be positive and that

$$-\frac{K_1 T}{mV}\left\{\left[-K_3 - \frac{K_2 T}{mV}\right]\left[-K_4\right] + \frac{K_1 T}{mV}\right\} - \left[-K_4\right]^2\left[-\frac{K_0 T}{mV}\right] > 0$$

Either K_3 or K_2 can be omitted without causing instability. To simplify analysis omit K_2 which would probably be the more difficult term to apply in practice.

Since all values of K are normally negative it is evident that the first stability condition is satisfied. Rearranging the second condition and assuming negative values for K_0, K_1, etc gives:

$$\left|K_3\right| > \left|\frac{K_1}{K_4}\frac{T}{mV}\right| + \left|\frac{K_4}{K_1}K_0\right|$$

This indicates that K_3 should be large, and (if a value of K_0 is established) that $\frac{K_1}{K_4}$ should approach $\sqrt{\frac{mV}{T}K_0}$

FORM 25-S-1
(REV. 8-43)

PREPARED BY: E. W. Graham

DATE: May 2, 1946   SANTA MONICA   PLANT

TITLE: PRELIMINARY DESIGN OF SATELLITE VEHICLE

PAGE: 3H

MODEL: 1033

REPORT NO. SM 11827

(B) Initial Stages

M = pitching moment

I = moment of inertia

T = thrust

d = displacement of thrust axis from C.G.

ϵ = actual heading

δ = design heading as a predetermined function of time

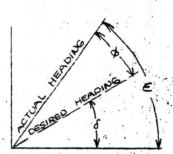

$$M = I\ddot{\epsilon}$$

Neglecting time lag let

$$\frac{M}{I} = K_0\int(\epsilon - \delta)dt + K_1(\epsilon - \delta) + K_2(\dot{\epsilon} - \dot{\delta}) + \ddot{\delta} + \frac{Td}{I} = \ddot{\epsilon}$$

Where K_1 and K_2 represent magnitudes of artificially applied restoring and damping moments, and K_0 is an integral term necessary for approaching the exact desired heading when eccentric thrust is present. The term $\ddot{\delta}$ is artificially applied as a predetermined function of time.

Differentiating the equation

$$K_0(\epsilon - \delta) + K_1(\dot{\epsilon} - \dot{\delta}) + K_2(\ddot{\epsilon} - \ddot{\delta}) - (\dddot{\epsilon} - \dddot{\delta}) = 0$$

$$\text{OR} \quad \dddot{\phi} + \ddot{\phi}(-K_2) + \dot{\phi}(-K_1) + \phi(-K_0) = 0$$

where $\phi = \epsilon - \delta$

For stability the coefficients of ϕ, $\dot{\phi}$ etc must all be positive, and since K_0, K_1 and K_2 are normally negative this condition is satisfied

Also for stability

$$(-K_2)(-K_1) = -(-K_0) > 0$$

$$\text{OR} \quad |K_2 K_1| > |K_0|$$

(C) General

Future investigations of stability might well be approached in a somewhat different manner. Since these systems are stabilized by entirely artificial means it would seem desirable to stipulate the type of motion desired and the damping and then determine the necessary values of K_1 K_2 etc to accomplish this. This approach should be mathematically simpler also.